C00 21919984
EDINBURGH CITY LIBRARIES

D0310577

The FOX and the ORCHID

For all those wanting to preserve the British countryside;
its traditions, its people and its wildlife.

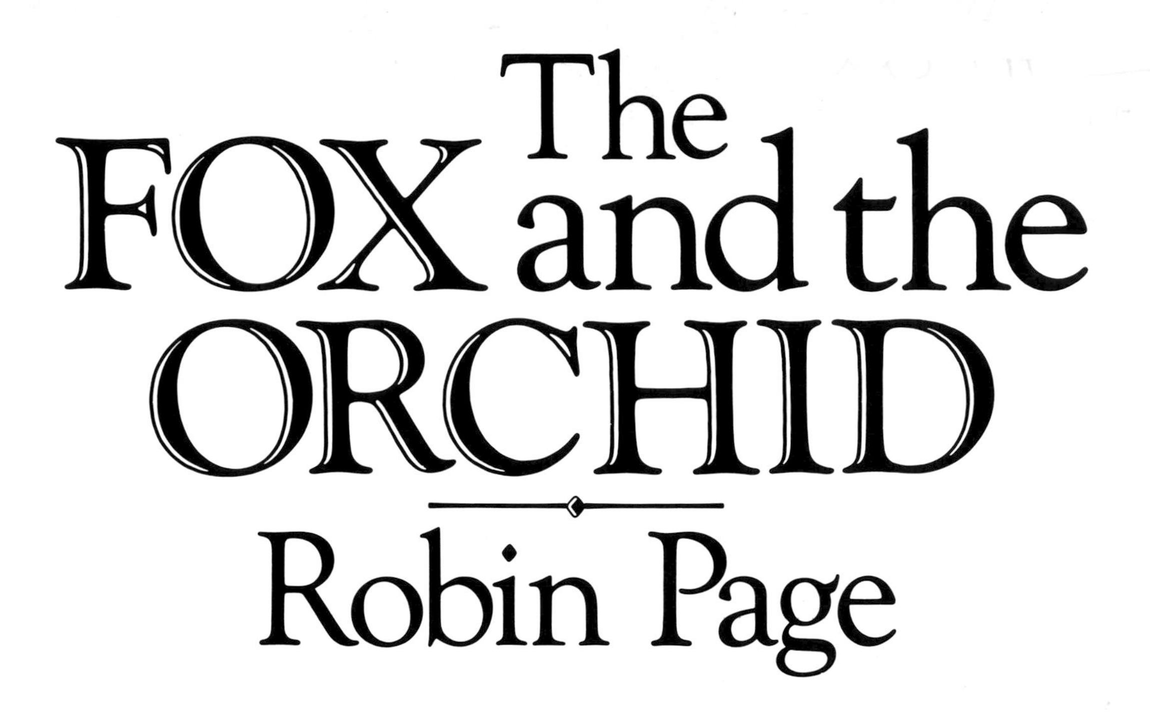

The FOX and the ORCHID

Robin Page

Country Sports and the Countryside

Quiller Press

By the same author:
The Benefits Racket
Down Among the Dossers
The Decline of an English Village
Weather Forecasting the Country Way
The Hunter and the Hunted
Cures and Remedies the Country Way
Animal Cures the Country Way
Weeds the Country Way
The Journal of a Country Parish
The Country Way of Love
Journeys into Britain
The Wildlife of the Royal Estates
Dust in a Dark Continent (Africa)
A Fox's Tale
Gardening the Country Way
A Peasant's Diary
Gone to the Dogs
Vokal Yokel
Carry on Farming
The Hunting Gene
One Man went to Mow
Dust in a Dark Continent (Africa)
The Great British Butterfly Safari
Children's Books:
How the Heron got Long Legs
Why the Rabbit Stamps its Foot
How the Hedgehog Got its Prickles
Why the Reindeer has a Velvet Nose
How the Fox Got its Pointed Nose

Or it may be said that hunting is ever a love-affair. The hunter is in love with the game, real hunters are the true animal lovers.

Karen Blixen — *Out of Africa*

First published in the UK in 1987
by Quiller Publishing Ltd
Second edition 2004

British Library Cataloguing-in-Publication Data
A catalogue record for this book
is available from the British Library

ISBN 1 904057 32 2

Printed in Singapore by Kyodo Printing Co (S'pore) Pte, Ltd

Quiller Press
an imprint of Quiller Publishing Ltd
Wykey House, Wykey,
Shrewsbury, SY4 1JA, England
E-mail: info@quillerbooks.com
Website: www.swanhillbooks.com

FRONTISPIECE:
The common spotted orchid by Robin Page

Contents

INTRODUCTION

to the Second Edition of *The Fox and the Orchid*

When asked if I would update a new edition of *The Fox and the Orchid* I readily agreed. For this I expected to write a completely new Foreword – but on re-reading the original – after sixteen years – I decided to leave it untouched. Despite the passage of time it still puts clearly my position on country sports and it explains the important link between country sports and conservation in the general countryside.

It shows in modern political parlance that country sports are 'green' – although few political and conservation 'greenies' are willing to put their heads above the parapet of political correctness to be counted; it would seem that fad and fashion are now more important than fact and straight forward honesty.

But in sixteen years there have been other changes – social, political and economic. Britain has become ever more detached from nature – suburban and urban Britain is creating semi-detached and disconnected minds in which the world of bunny huggers and anthropomorphic television is luring people away from the realities of wildlife and conservation. At the same time the supermarket has broken the connection of season, soil and culture from the production of food.

Partly as a result of this disconnection, rural Britain has found itself under unprecedented political and media attack. While Britain has been fighting for 'freedom and democracy' and the rights of minorities in Bosnia, Afghanistan and Iraq, Britain's own rural people – a minority culture in multi-cultural Britain – have seen farming brought to its knees; rural communities breaking up and hunting under attack in the House of Commons. At the time of writing hundreds of farmers and farm workers are leaving the land every week because of economic collapse and in 2002 the official figure was 1000 every week; in addition there is a farming suicide every six days. With such a disaster taking place in any other part of British Society there would be a mixture of concern and outrage; because it is taking place across rural Britain in an urban dominated society it is ignored. Democracy? What democracy – this tragedy is about domination, prejudice and discrimination.

Of course the political attacks are aided by a seemingly inexhaustible supply of cuddly bunny money with the Labour Party pocketing well in excess of £1m from anti-hunting organisations over recent years. Under our First World political system such payments are called 'donations': under Third World systems they are called 'bribes' and described as 'buying political favours'. Astonishingly it has been estimated that in the last five years alone, over £30m has been spent on the anti hunting, propaganda war by the major 'animal welfare' organisations. Significantly, I believe, a number of serving MPs have refused to tell me how much money they have received during this time to keep the political pot boiling. Just think what impact this money would have made to real conservation. With most serious conservation bodies desperate for money and The Countryside Restoration Trust urgently wanting £1.5m.

Changes have come too in my personal life. I am no longer the 'ol' boy' on the family farm as my brother is now part of the statistics of those leaving the land: disheartened he gave up – and so I am now the 'farmer' and am determined not to give up – although I admit that my pen allows me to be independent from family farming's subsistence level income.

In 1993 I helped to launch The Countryside Restoration Trust – Britain's first farming and wildlife charity – farming its own land. It has been an exciting, white knuckle ride and from no members, no money and no land we have climbed to nearly 1000 acres, three farms, one in Cambridgeshire and two in Herefordshire, a bluebell wood in Yorkshire and a field and spinney in Sussex. We have raised nearly £3m: we have over 6000 members, and more alarmingly we have a bank overdraft of £870,000 – the white knuckle ride continues. At Lark Rise Farm in Cambridgeshire – the return of wildlife has been spectacular – simply by tweaking the farming system with the addition of beetle banks, grass margins, smaller fields, hay meadows and hedges, otters and barn owls have returned – the barn owl after a gap of 45 years. The brown hare population is like it was in the 'fifties: we have possibly the highest skylark density in East Anglia and we have yellowhammers, English partridges, reed buntings, corn buntings, harvest mice, grass snakes and butterflies and bumble bees galore. It has been a wonderful and positive time and our tenant farmer Tim Scott farms with a smile on his face.

We have shown that with a mixture of care and concern wildlife will return. But what this book shows conclusively is that where hunting, shooting and fishing take place – the extra habitats created for country sports have brought the wildlife back already: so what benefits the fox and the pheasant also benefits the barn owl, the otter and the bee orchid – yes, The Fox and the Orchid.

In addition to all this, in the year 2000 I wrote another book on hunting *The Hunting Gene* – a successful book selling over 10,000 copies. Ironically I had to take a huge financial gamble and publish it myself as no mainstream publisher would take it on – deeming it to be 'politically incorrect' – anther blow to Britain's diminishing reputation for 'freedom'.

The book was important for me as it took me beyond the conservation and ethical issues. For the first time it took me into the world of 'Fell Packs' and 'Foot Packs' – it took me into the cultural world of hunting – a culture involving communities and all sections of society – a jigsaw of farming, dogs, shepherds, foxes, followers, cattle, wildlife, socials, horses, dinners, balls, point to points, singing, working, crying, laughing – a thriving, living culture under attack – incredible!

I finished the book as Lord Burns was starting his Inquiry into 'Hunting with Dogs' for the Government. Our paths crossed several times and he seemed a fair and affable man. His work and that of his four colleagues was presented to Parliament in June 2000 as the 'Report of the Committee of Inquiry into Hunting with Dogs in England and Wales'. The Report, although deliberately limited by the Government in its objectives, is good, and I refer to it in the main text here as 'The Burns Report'.

I still do not hunt, shoot or fish – although I followed hounds journalistically several times for *The Hunting Gene*. I have also 'walked' a foxhound for the Cambridgeshire hunt. Called 'Corset' she was a wonderful, politically incorrect dog – ending up at 70lb and wanting to sit on laps, she tried to sit on my lap one day as I was driving along the M11 – and she loved drinking from the lavatory.

A year ago I did actually break my rule and hunt for fun when Alun Michael, the singularly unimpressive Minster for Rural Affairs announced yet another anti-hunting Bill for the House of Commons. I went to ride with the Devon and Somerset Staghounds as an instant act of solidarity with the good and hard pressed people of Exmoor. Among his various other titles this weak politician is also called Minister for the Horse and Minister for Urban Quality of Life. So the satirical worlds of George Orwell settle into the real world of Blair's Britain and Blair's government.

There have been other sadder changes too – the inevitable ones that the passing of time bring. Many who contributed to the first edition of *The Fox and the Orchid* are still in full swing while others such as Lt Col Murray-Smith, Lord Home, the former Prime Minster, and the composer John Hotchis have moved to greener, high altitude pastures. Gordon Beningfield, the artist and passionate conservationist has also gone – cancer took him after a courageous fight. He was my closest friend and so important to the CRT; I still miss him. He did not hunt, shot rarely and fished occasionally – yet he bravely defended country sports for their importance to conservation and for their contribution to rural culture. I have retained all these contributions as the validity of their arguments is the same today as it was yesterday – and, more importantly it will be the same tomorrow.

So here is the new edition of *The Fox and the Orchid*. I hope people will read it with an open mind – remembering that the British countryside is a living, breathing entity in its own right – it is not just a Theme Park – the political plaything of urban politicians.

FOREWORD

This is the book everyone who loves the British countryside should read. Of course it will be bought and read mostly by country sportsmen and women because it defends their passions and their activities so very well. But it should really be read by those who have doubts about the morality and benefits of country sports, for no other book captures the strange love affair that people who hunt and shoot and fish have with their prey. And no other book is as persuasive in putting across all the well rehearsed reasons why they are 'a good thing'.

Packed with information and sometimes startling facts about the peculiar workings of the countryside and the plethora of symbiotic relationships which hold it all together, it should be on the school curriculum for all urban children, but this is most unlikely to happen in today's hostile climate to all things rural. A pity, because even the most cynical reader is likely to be swayed by Robin's eloquent writing. No one can doubt the sincerity of his love for nature, and his ability to capture the beauty of our landscape and its wildlife is what lingers in the memory long after this book has been read. And, unlike most books about the countryside, this one demands to be read to the end – and then kept at the bedside for further enjoyment.

The Fox and the Orchid has always been one of my favourites and it is very good to see it being republished in a new edition, completely updated to reflect all the recent dramatic changes that have affected so many of us.

Robin Hanbury-Tenison

Explorer, Conservationist, Broadcaster, Film Maker, Author, Lecturer, Campaigner, Farmer and, from 1995 to 1998, Chief Executive of the British Field Sports Society/ Countryside Alliance.

Introduction

The Fox and the Orchid concerns Britain – my country – whose people, traditions and countryside are more important to me than those of any other part of the globe; for although we tend to equate the destruction of wildlife and the disappearance of old customs with the developing world, yet here in Britain we are still damaging and destroying our own unique landscapes under the guise of 'progress' and 'agricultural efficiency'. As a direct consequence much wildlife is under great pressure as its habitat diminishes, and some species have even disappeared.

This book is about parts of our own rural tradition – those country sports involving hunting, shooting and fishing. By their opponents they are known as 'blood sports', but by those who participate they are referred to as 'field sports'. 'Country sports', however, seems to be the most accurate description, as they take place mainly in the country, and for many generations they have played an important part in the shaping of the British countryside.

Now, country sports are under political threat; it has become fashionable to be opposed to 'blood sports' on the grounds that they are 'cruel', 'barbaric' and a 'threat to wildlife'. Such an appeal is attractive, for most people are naturally and emotionally opposed to cruelty, as they are against the destruction of nature. Politically it has another bonus, in that the sports themselves are assumed, quite wrongly, to be the preserve of the 'upper classes' and so those who oppose them can wage two popular wars at once – against both class and animal cruelty. As a result the movement to ban country sports has undoubtedly gained ground over recent years; it has also grown in popularity because the vast majority of people in Britain today live in towns and they do not actually understand what country sports involve. They will condemn foxhunting without ever having seen a fox or knowing details of its natural history, and they will ridicule grouse shooting as a trivial aristocratic pursuit, failing to realise that without the shooters both the grouse and the moorland it inhabits would be under real and immediate threat.

The purpose of this book, therefore, is to explain simply and briefly the role of hunting, shooting and fishing, and, more importantly, to show the vital links between country sports and conservation for, contrary to the popular view, they are of great benefit to wildlife and constitute an important element in the preservation of the traditional British landscape. The contribution made by country sports to conservation was shown to me exactly ten years ago when I wrote *The Hunter and the Hunted.* For that book I learnt to ride, and witnessed every sport at close quarters, in order to evaluate its place in country life, and consider whether it was acceptable on ethical grounds. This last point was extremely important to me as I have always abhorred cruelty and as a practising Christian I would be opposed to pastimes based on the needless suffering of animals simply for the pleasure of sportsmen.

However, like others before me, I found little cruelty; what I did find were well conducted traditional country activities. Where I found cruelty I said so; I condemned the use of self-lock snares (now banned); I opposed otter hunting (now banned) and I also criticised some aspects ofwildfowling and fishing that have remained unchanged and are discussed later.

Finally, I must make my own position on country sports clear. Although I am extremely interested in them, because of their place in country life and the contribution they have made and are still making to the countryside, I do not participate in any myself. I have a gun, but have not used it for two years; the last occasion was to stop a pigeon eating my cabbages. I did pick it up in anger this year on noticing a magpie systematically searching for young blackbirds in my garden hedge, but on seeing the gun it quickly made off.

The reason that I do not hunt, shoot or fish is simply that I like watching birds and animals and have no desire to kill them. However, I do accept that other people have different views and feelings – as the instincts of the hunter are quite natural ones -and I also realise that some killing and 'management' of wildlife populations are inevitable in areas where man and wildlife share the same land.

On our small family farm we have kept hedgerows and grass meadows, because we like them and the wildlife they attract. Elsewhere however it has been different; the countryside has, in many places, been ravaged by the intensification and industrialisation of agriculture as the march of the prairie farmers goes on – at the direction of the E.U.'s Common Agricultural Policy. of the. Yet where a farmer hunts and wants cover for his foxes, or shoots and retains areas suitable for nesting pheasants, there the prairies end and the traditional countryside remains. My main purpose in writing *The Fox and the Orchid* is to show this real but often overlooked link, both in the past, and, more importantly, in the present.

R.P.

Chapter 1

HOME GROUND

Over recent years successful campaigns have been organised in Britain to help save the tiger, the panda, the black rhinoceros, tropical rain forests and the Serengeti from extinction and destruction. Such efforts are to be commended, but it is strange to record that at the same time as people have become more aware of the dangers facing various exotic species and distant places, much of our own British countryside – and the native flora and fauna which depends on it, has been devastated. It is true that organisations such as the Royal Society for the Protection of Birds and various voluntary and official wildlife organisations have created numerous reserves and refuges throughout the country, but simultaneously much of the general countryside – particularly in lowland Britain, has been damaged almost beyond recall.

There are several culprits, all of whom have sheltered behind words such as 'progress', 'development' 'efficiency' and 'competition on the world market'; short term political gains and private and corporate profit have been put before long term aims, the needs of future generations and our responsibilities for preserving our natural heritage. Developers and planners have covered thousands of acres with sprawling estates, motorways and airports; sometimes in rural areas protected by the Green Belt – a sacrosanct area supposedly safe from development.

At the same time many farmers have responded to the political demands for 'cheap food' and maximum production by turning huge tracts of once attractive countryside into a series of featureless prairies, where chemicals, and insensitivity reign supreme. Those really responsible, however, are the politicians, both national and local, who actually created and allowed these policies of devastation. Even with their own land, local councils are often guilty of acts of environmental hooliganism and they, together with Parliament, have totally failed to control the activities of the country's Water Authorities, (now after several name changes called The Environment Agency – a new organisation with old habits) who in just twenty-five years have destroyed most of the habitat suitable for the otter, one of Britain's most attractive animals, and have turned countless once attractive rivers and streams into featureless drains.

Ironically, and sometimes cynically, politicians are realising that there is popularity – and votes – to be gained from supporting 'green' policies and they are taking a mainly superficial interest in 'conservation'. Consequently many are proclaiming that acid rain, pollution, prairie farms and the threat of wildlife extinction are the 'great issues' of our time, despite the fact that their own policies created the problems in the first place. Often, in their desire to be seen as 'green' and environmentally sound, they advocate plans just as absurd and potentially damaging as those that have gone before, based on ignorance, emotion, prejudice and expediency; proposals to outlaw 'blood sports' and to create 'sustainable communities' (allowing wholesale development) – fall into these categories.

The figures for change speak for themselves. They show that without a doubt the face of rural Britain has been almost totally transformed since the end of the Second

A view of what is commonly called 'prairie farming', in the author's neighbourhood together with the dishes of Cambridge University's famous Radio Telescope.

World War in 1945. Intensive agriculture has altered the lowlands, changing a patchwork of meadows, woods and arable fields, divided by hedgerows, into huge blocks of intensive, chemical-based crop production. In many areas landscapes that were once aesthetically pleasing and rich in many forms of wildlife have been turned into uniform factory farmland, where the wildlife remnants have to struggle to survive. Upland regions have not escaped, for there huge blocks of alien fast growing conifer trees have been planted, converting wildlife-rich areas of moorland and ancient forests into ecological deserts. The speed and degree of change was confirmed by the Nature Conservancy Council in 1977 when it wrote in its Annual Report: 'The rate and extent of change during the last thirty-five years have been greater than at any similar length of time in history.'

The facts are frightening; since 1945 Britain has lost over 150,000 miles of hedgerow. This is the equivalent of a hedge passing round the equator over six times and represents more than a quarter of our hedges. Between 1946 and 1970 Norfolk lost half its hedges – 8000 miles, while Huntingdonshire lost 90%, or 5000 miles during the same period. It was part of an agricultural revolution in which more powerful and efficient machines required larger fields, and where chemicals – fertilisers, pesticides and insecticides – were

A mature hedgerow – bustling with wildlife of all kinds.

used in large quantities to ensure maximum returns. In this revolution, arable agriculture, and cereal monoculture in particular, ousted mixed farming in much of lowland Britain, with fields of grassland and cattle being replaced by acres of financially attractive winter sown cereals. Ninety-five per cent of Britain's traditional lowland hay meadows went during this time as well as 80% of chalk and limestone grassland. Many of the old hay and water meadows were drained, to allow the change, while the free draining chalklands offered no resistance to the plough. The old Dorset grassland was particularly rich, containing over twenty species of grasses, 120 flowering plants and a wealth of butterflies and insects. One quarter of Dorset's downland was ploughed between 1967-72 and half of Wiltshire's downland experienced a similar fate between 1937-71. Dorset's lowland heaths have also declined, by 42% since 1960, while those in Suffolk have virtually disappeared. The change is not limited to the south, for most of Lancashire's coastal bogland has been drained out of existence, too.

Land not suitable for intensive agricultural production is often considered ideal for forestry and over the last ten years 750 square miles of Britain has been covered with fast growing, alien softwoods. Despite this the Forestry Commission hopes to plant another four million acres with trees, 97% of which will be conifers. Pine or fir plantations are virtually useless as far as wildlife is concerned; the trees themselves offer suitable habitat for a handful of small birds and little more, and they are planted so close together that after a few years of growth all the light is kept out and the forest floor is covered only with needles. To make matters worse, it is now widely accepted that the large areas of conifers are helping to hasten the acidification of our rivers and streams, adding to the problem of acid rain.

The silence in these plantations has an eerie, dead quality that can also be experienced in the forests of Scandinavia and Northern Europe. Recently a group of Swedes walked through an ancient English broad-leafed wood during springtime. They were astonished at the birdsong, the blossom, and the many woodland flowers; it was the first time they had seen a wood bursting and throbbing with life.

But although alien woods have been created, our own native woodlands have suffered badly. In 1945 there were about 11 million acres of semi-natural woodland in Britain. Since then 30% of the area has been covered with conifer plantations; 10% has been cleared; 50% has been neglected and only 10% is managed in the old traditional way. To make matters worse it has been estimated that by the year 2025 there will be no natural, or semi-natural woodland left in the British Isles outside nature reserves.

Since the end of World War Two over 37,000 acres each year have been lost to development and afforestation – the equivalent of 50 football pitches a day. This represents a vast expenditure in financial terms, but the cost is far greater than simply money, for it has also meant a drastic reduction in our non-renewable (natural) resources.

The cost of each new motorway is not just measured in millions of pounds per mile it is 42 acres of good farmland or wildlife habitat for every mile – the equivalent of 35 football pitches, plus 10,000 tons of aggregates per mile, including quantities of sand, gravel and hard core. Development does not take place on inland sites alone, however, for two thirds of the natural coastline of England and Wales has been destroyed.

To future generations all this could be vital, particularly when it is remembered that according to numerous estimates the world population will double within twenty-five years, and that by the year 2020 there will be a global crisis of over-population, starvation, increasing pollution and the exhaustion of many of the natural resources on which we currently depend. Seen in this context the current food surpluses in the West, that anger politicians so much, could be very short-term – but then regrettably British politics is also a very short-term and imprecise science, looking only as far forward as the next general election.

Unfortunately this catalogue of 'achievement', gained through 'progress' can be clearly measured in wildlife terms; the barn owl has disappeared from vast tracts of land, the corncrake can only be found on the Western Isles of Scotland; the large blue butterfly disappeared and has had to be re-introduced and ten species of plant have become extinct since 1930. Although various conservation bodies are aware of the problems and dangers, many more species remain under threat. Several types of bat are endangered, as well as the smooth snake, the natterjack toad and the sand lizard. Twenty-seven species of wetland plant are threatened; twenty-two are rare and twenty-eight are scarce. Once common flowers such as cowslips, orchids and fritillaries are uncommon and have to be given special protection where they occur. Even the 'common frog' is today uncommon; in Huntingdonshire its numbers plummeted 95% between the 1930s and the 1970s.

In addition it is thought that three or four of our forty-three dragonflies have become extinct since 1953, with six more vulnerable. Ten of our fifty-five species of butterfly could follow the large blue into extinction, and another thirty-five are in decline. A similar picture could be painted of spiders, bumblebees and moths. Birds have also been hit, with at least thirty-six species declining significantly in the last thirty-five years, including the barn owl, bittern, nightingale, kingfisher, skylark, cornbunting, reed bunting and lapwing.

In theory the Government's Wildlife and Countryside Act of 1981 gave protection to the countryside, and over 4000 SSSIs – Sites of Special Scientific Interest — were re-scheduled. Their protection was, in the first instance, voluntary with certain statutory safeguards. Despite this over 4% are damaged each year and Ministers of State still frequently rule in favour of farmers, developers, industrialists and local authorities who want other uses for them.

To me the realities of change and 'progress' cause genuine and deep feelings of personal sadness. I was born on a farm in a small Cambridgeshire village in 1943 (next door to where I still live today) when the age of accelerated change was about to begin. On our farm we still had two cart-horses, as well as one tractor, and the farming was mixed – a mixture of cereals, fodder crops, grass and fields left fallow. Each farm in the parish had its own herd of dairy cows, with most being dairy shorthorns.

Nearly all the fields and meadows were small, divided by hedges, with hedgerow trees, and there were several small spinneys and copses. A brook marked the southern boundary of the parish, and each winter after heavy rain it would flood; in sunlight from the old railway bridge, under which steam trains still passed, it seemed like a wide silver ribbon winding through water-meadows to the Cam.

Every spring we would see hares in their March madness and coveys of partridges

The natterjack toad – one of Britain's endangered species.

would feed in the autumn stubbles. The corn was cut by binder and when the stacks were threshed mice and rats would flee in disarray, with some falling prey to the talons of hovering kestrels. In the brook meadows were several large rabbit warrens, and among the grasses, cowslips, buttercups and daisies grew in profusion. Late on summer evenings barn owls quartered the grassland, grey, ghost-like shadows in the mist, and each July we would pick armfuls of cornflowers, knapweed, scabious, poppies and assorted grasses, for the village flower-show's wildflower competition.

Near the surface of the deep pools of the brook, large pike would bask, and we would catch roach, perch and dace, sometimes taking them home to eat. Other fish escaped, including one 'monster' eel whose size increased from year to year. I often sat on the brook bank without my fishing rod, simply watching the damsel flies settling on the water lilies, or enjoying the sight of the 'yellow flags'* on flower**, and listening to the flow of water over the shallows. Many other creatures lived along the brook: moorhens nested in the reeds, 'water rats' chewed busily, and every year the otter hunters followed scent from the River Cam into the lower reaches of the brook. An

---

*yellow iris
** 'on flower' is a piece of Cambridgeshire dialect and it still sounds more suitable to me than 'in flower' which some people claim to be grammatically correct.

How it was once – a drawing by Gordon Beningfield

old man fished every summer and he told stories of watching otter cubs playing – sliding and swimming with their mother. Their holt was in the roots of an old bank-side hawthorn tree.

Most of the men in the village worked on the farms and we had Jim, a First World War veteran with only one eye, from a neighbouring village. He loved his work – ploughing, drilling (sowing) and harvesting. He understood the land, too, as well as the creatures that shared the fields with him. As the corn was cut nothing would give him more pleasure than seeing a family of pheasants or a covey of partridges, and if a nest was destroyed accidentally at hay time he would rescue the undamaged eggs and put them under a broody hen.

In addition to the cart-horses and cows, there were other animals around the farmyard itself. Pigs, geese, muscovy ducks, bantams, free-range hens and an old grey donkey that we would sometimes ride. Dogs and cats were always present, in and out of the house, and for several years a hen, Henrietta, would walk into the living room to check for scraps. Often pets were very much part of the family – creatures brought in from the wild – rabbits, jackdaws and a tawny owl chick that had fallen out of its nest. Newts and frogs also appeared from time to time, in direct contrast to the tortoise who would regularly manage to disappear.

Change came slowly at first, but gradually accelerated until by the late sixties and early seventies it had reached a stage of near frenzy. Horses were replaced by tractors, and then the early tractors were replaced by larger and more powerful models which allowed the work to be done faster and more efficiently. It was in the 'fifties that for the first time my father began to use artificial fertilisers and sprays, and we have continued to do so ever since. As yields began to increase with a more scientific approach, wheat and barley were grown more frequently, in preference to other crops, as they became more attractive financially. The change to cereals was a real agricultural revolution; it meant that all the local farms, except our own, got rid of their dairy herds and most of their grass meadows were ploughed. Then, with the help of government grants, some farmers began uprooting their hedgerows to make larger fields, more suited to the use of modern tractors, reversible ploughs, large sprayers and wide combine-harvesters.

Cereal growing was more attractive in other ways, too, for whereas livestock demanded attention seven days a week, cereal production permitted a five-day week for most of the year, the same as bank clerks and factory workers. As more machinery was used, fewer men were required to work on the land and many left to get better paid and less arduous work in the town. As they became commuters, others, 'newcomers', already commuting, appeared in the village. New middle-class estates were built where most of the men had salaried office jobs in Cambridge; consequently a new type of countryman and country living came into the village.

One of the new estates was built on the site of a grove of high elm trees, where each spring rooks nested in the top-most branches. The trees were felled when the rookery was full of young rooks unable to fly or fend for themselves – it was carnage. Another small estate was built in an ancient grass meadow where, every year, until their site was invaded, jackdaws nested in the holes of old doddled* elms.

By this time other wildlife had already disappeared. It was in the 'fifties that the evil disease of myxomatosis was deliberately introduced into Britain because of the economic damage being done by rabbits. Within a few months of the first sick rabbits appearing, all the warrens were empty and it seemed that the rabbit would become extinct.

---

*pollarded

The next nail in the parish's wildlife coffin came in the cold winter of 1962-63* when snow and ice lasted for months instead of weeks and did not let up until April. When spring finally arrived several birds had disappeared completely – the magpie, green woodpecker, kestrel, sparrow hawk, kingfisher, and all the owls, barn, tawny and little. At first it was thought that the cold had simply killed them, but later it was revealed that nearly all over the country the situation was the same and the freeze-up had been the final act of a tragedy. The birds had already been weak because of the increased use of DDT based pesticides, with dieldrin being the most toxic substance. In addition to weakness, breeding success and fertility had also been adversely affected; then came the cold, which killed off thousands of poisoned birds. The chemicals were banned and a national wildlife disaster was only just averted. Since then magpies, kingfishers, kestrels, tawny owls, green woodpeckers and sparrowhawks have all returned – slowly at first – but the little owl remains scarce

In 1972 yet another disaster occurred, when the Great Ouse River Board (now the Environment Agency) decided to 'clean out' the brook. In the course of a few weeks mechanical diggers gouged out the brook's bed; trees on its banks were felled and thickets cleared. What had been a rich wildlife haven became a large open drain, similar to those found in the Fens, with a uniform rate of flow and no wildlife cover. The hawthorn otter holt was uprooted; flowering rush, marsh marigolds, yellow irises and water lilies disappeared, pike slowly starved to death, and the reed bunting became a rarity.

The water table was lowered by at least five feet, draining a pond where frogs had bred for generations, and the water meadows lost all their water. Astonishingly the engineers started their work upstream and worked down – totally reversing normal drainage procedures and possibly causing the brook to be dug out deeper than intended. But although expressing surprise, both the Department of the Environment and the Ministry of Agriculture stated that they could do nothing, as the water authority was a fully autonomous body. The conservation bodies were in a difficult position, for they had other areas where they needed the water authority's help and so they had to avoid alientating those concerned; as a result their public objections were muted, while their private, 'off the record' anger was unprintable.

The cleaning out of the brook meant that more land could be ploughed and several neighbouring farmers ripped out the long, high hedges of their old water meadows to incorporate them into their arable fields. It is ironic; at the time I complained that the work would simply lead to unwanted grain and that the financial cost to get fifty or sixty acres into cereal production was not worthwhile. Since then Europe has produced a 'grain mountain' *leading to land being taken out of production through 'set-aside'.* Unfortunately water meadows cannot be re-created simply by a change in government

*see also *The Decline of an English Village*

The Bourn Brook immediately after being cleaned out by the Anglian Water Authority – lifeless and angular.

policy; it will be many years before nature can repair the damage done and it is most unlikely that away from land owned by the Countryside Restoration Trust the old hedgerows will ever be replaced.

We were the only farm in the parish to remain virtually unchanged; my father keeping all his hedgerows and his dairy cows, believing that mixed farming with a rotation of crops was healthier for the land. At the same time we started a long-running battle with the water authority to stop their workmen cutting all the natural re-growth along the brook banks every year and to reinstate our frog pond so that we could re-introduce frogs. It took twelve years of agitation, and the passing of the Wildlife and Countryside Act which required water authorities to take conservation interests into account, before the pond was partially restored and the brook banks were left almost untouched. Now, bank-side trees are again growing and with a little help, a few clumps of yellow irises and water lilies have returned, but marsh marigolds are still absent

Once the pond had regained some water, one clump of frog spawn appeared in its first spring, and we introduced additional spawn, tadpoles and two frogs from elsewhere. Now every spring numerous frogs appear at spawning time and a brief glimpse is given of true wetland life. Although the return of the frogs has given us pleasure it seems remarkable that it took years of fighting bureaucracy to restore just one small area for wildlife.

A motorway has now been added to the list of damage,* slicing through the countryside between the village and Cambridge; it skirts a small hill, where I *once* watched badgers; *forget the badgers – it has made life so much* easier for people to commute to London and beyond. The road was not wanted, but its construction was inevitable. One day, several years ago, a local farmer (a sensible one) noticed surveyors hammering white pegs into one of his fields. When he asked what they were doing he was informed that they were checking the site of a proposed motorway and added that if it came, Stansted would become London's Third Airport, as the two schemes were linked. At the public enquiry into the Cambridge Western By-pass (the M11 Motorway) the link with Stansted was not mentioned, although it remained clearly marked on the planners' drawing boards. Conseqently when the government ignored the findings of various public enquiries to announce that Stansted would become London's Third Airport I was not surprised in the slightest.

At about the same time Dutch Elm disease added one more despondent chapter to the parish; the great old trees of a nearby spinney slowly died and the whine of chain saws replaced wind in the leaves. *In 1980* yet another change came; because of the financial pressures on small family farms we accepted a financial inducement from the Ministry of Agriculture to get out of dairy cows, to help reduce the size of the European butter mountain. It was a miserable time and I stayed away from the farmyard when the cattle

---

*see *The Journal of a Country Parish* O.U.P.

lorries arrived to be loaded up. Now we too grow mainly cereals, contributing to the European grain mountain instead of the butter mountain, and grass does not feature in our crop rotation. But unlike most of our neighbours we have kept almost all our grass meadows by the brook, where beef cows and their calves run in the summer, and we bale much of our cereal straw for feed and bedding for the cattle in winter. In addition we have kept all our hedges and some of them we have 'let go' to grow almost wild. We have kept them quite simply because we like them, both for ourselves and for our wildlife. They provide cover and roosting places for pheasants; hunting areas for foxes; blossom for butterflies and bees, as well as a natural harvest for birds and people — blackberries, sloes, hips, haws, wild crab apples, hops and spindle berries.

One of our immediate neighbours also farms sympathetically, planting trees and trying to care for his hedges and his land, but beyond, almost featureless prairies stretch for acres on end. The wild flowers have been sprayed away, ditch banks are flail-mowed for tidiness and the remnant hedgerows are trimmed back so much that they are of no use to wildlife. Often the hedge-cutting is done in late spring, when there is a lull in the land work, just when the wild birds are nesting and feeding their young. A few farms in neighbouring parishes have resisted this trend; they are the ones where sporting interests, shooting or hunting, are continued, and hedges, trees and wild areas are still wanted.

Although we have tried to resist the harmful aspects of farming fashion, yet more pressure and change is on the horizon and approaching fast. Just as northern England succumbed to 'growth' and almost uncontrolled development during the Industrial Revolution, so developers, planners and the government are doing the same to East

Removing a hedgerow to make way for a dual carriageway in Wales.

Anglia during the present technological revolution, with the M11 motorway, Stansted and high technology industries all adding to the pressures.

For the politicians the short-term attractions of this folly are almost irresistible, allowing them to use much favoured words such as 'growth', 'employment' and 'opportunity'. But by the time the local economic bubble bursts, as it always does, the harm will have been done and the countryside and its wildlife will have contracted still further. It is in this context of continuing rapid change, both nationally and locally, that country sports should be viewed. To make matters worse the Government – now dominated by urban Members of Parliament has announced (in 2003) a massive housing development of 250,000 houses for South East England, of which 50,000 are scheduled for South Cambridgeshire. It is environmental madness – illiteracy of an astonishing kind – accompanied by the announcement of seven new major airport runways around London, spreading the pollution of noise, light and air. One of these is said to be needed at Stansted – needed like a candle and a box of matches in a firework factory.

The facts are that Britain's population is growing by just 0.18% a year – if economic migrants (called asylum seekers by the politically correct) are controlled, this figure should fall to 0.04% by 2025 So why are new houses proposed on a massive scale when the population is almost stable. The answer is simple – most of the new high-tech jobs are in Southern England and so the Government is allowing a gigantic flow of people to jobs, regardless of the effects on the environment and communities – rather than orchestrating responsible policies to direct jobs to people.

This and other woodcuts in the book are by the legendary Thomas Bewick.

Chapter 2

The Fox and the Orchid

The country sport with which most people are partially familiar is foxhunting. It is not the oldest form of hunting with hounds, but the most aesthetically pleasing. There can be few households in the country that do not have at least one Christmas card each year depicting horsemen, hounds and a fox, streaming across a winter landscape. The landscape is always one of trees and hedgerows, the country of post enclosure Britain – the traditional and attractive countryside normally associated with foxhunting. Manufacturers would sell few cards showing a hunt over a flat, treeless East Anglian prairie.

The oldest hunting sport is possibly the pursuit of the hare, but in the eighteenth century foxhunting suddenly gained popularity and became the sport of the gentry. The midlands housed the great foxhunting countries of the Quorn, Belvoir, Cottesmore, Pytchley, Fernie and Fitzwilliam hunts, with Melton Mowbray being the fashionable centre. At that time there was no question of foxhunting being carried out to control the fox – indeed foxes to chase were sometimes so scarce that additional animals had to be imported from France and released. One of the great appeals was – and still is – the thrill of the chase, best summed up by Jorrocks, that wonderful character created by R. S. Surtees: ''Unting is all that's worth living for – all time is lost wot is not spent in 'unting – it is like the hair we breathe – if we have it not we die – it's the sport of kings, the image of war without its guilt, and only five and twenty per cent of its danger.'

It is interesting to note that foxhunting's popularity soared at a time of change in the country. Before the enclosures the great appeal had been the exhilaration of galloping hard over miles of open and uninterrupted country, with hooves pounding

P. Aubertin's depiction of the Chase.

and mud flying. At first the new landscape after enclosure was difficult to hunt, but once suitable techniques had been developed, including jumping, many found the new even more exciting than the old.

To John Clare the Northamptonshire 'peasant poet' (although his village is now in Cambridgeshire), the change was tragic and in his fine book *The Making of the English Landscape*, Professor W. G. Hoskins quotes one of his poignant verses:

> Ye fields, ye scenes so clear to Lubin's eye,
> Ye meadow-blooms, ye pasture flowers farewell!
> Ye banish'd trees, ye make me deeply sigh –
> Inclosure came, and all your glories fell:
> E'en the old oak that crown'd yon rifled dell,
> Whose age had made it sacred to the view,
> Not long was left his children's fate to tell;
> Where ignorance and wealth and their course pursue,
> Each oak tree must tumble down – old "Lea-Close" adieu.

Just as modern foxhunters have tried to resist change by retaining as much of the old landscape as possible, so the hunts of Clare's time planted woods and spinneys as fox coverts, and many of them still exist today.

In view of the reputation of the fox and the attractiveness of much hunting country, the virulent opposition to foxhunting is not always easy to understand. Part of the problem is caused by the fact that we are an urban dominated society, which has been strengthened by those many people with urban and suburban backgrounds who now live in the country. They have taken their town views, values and prejudices with them, to exert change on the traditional structure of village life and thought.

Another reason is quite simple – that most people do not understand the fox or its natural history. They are affected by 'the cuddly bunny syndrome' and carry the image of the childhood fox with them, on into later life. They see Reynard as a soft, furry creature, much maligned – mischievous rather than sly or cruel. This view is obtained from several well loved story books; Brer Fox, in spite of his efforts to capture and eat Brer Rabbit, receives sympathy, as he is always outwitted – even humiliated – as he attempts to claim a dinner. Other foxes are more successful: one eats Chicken Licken and a whole host of assorted farmyard fowl after the rather dim little cockerel was hit on the head by an acorn; he came to the unscientific conclusion that the sky was falling down. Another double-crossed the Gingerbread man, when, instead of helping the unfortunate piece of neurotic confectionary to cross a river, Reynard gulped him down in mid-stream.

The Three Foxes created by A. A. Milne had no nasty traits at all and were distinctly lovable – 'Once upon a time there were three little foxes, Who didn't wear stockings and they didn't wear sockses'. They were rather odd, however, as they all had

The most famous of all the hunting hazards – the Wifsendine Brook (by H. Alkin).

'handkerchiefs to blow their noses, And they kept their handkerchiefs in cardboard boxes'.

In folklore too* the stories are all entertaining and the fox is not totally villainous. Although some tales show the fox as not too honest, he is, at the same time, a thoroughly lovable rogue. Hence the stories of how the fox acquired a white tip to his tail, and made hens fall from trees, are remembered with pleasure, as are those explaining how the fox rids itself of fleas and wears the flowers of foxgloves over its paws, to enable it to walk more silently. Aesop's fables are also well-loved – The Fox and the Crow, and the Fox and the Stork, but again his roguery is considered acceptable. I have added to this image of the loveable rogue by writing 'How the Fox Got its Pointed Nose' – which is not very Darwinian, or scientific – but could be true for those with imagination.

---

*see the Author's *A Fox's Tale*, published by Hodder and Stoughton in 1986, which covers in much greater detail the fox in fiction, folklore, fable and fact.

Consequently, with this background, the commonly held, long-term view of the fox is based on a mixture of fiction, folklore and fable, with a small amount of fact thrown in for good measure. The fox is seen as a much maligned animal who lives by eating voles, mice, rabbits and the occasional hen. When he does venture into the hen-house, it is simply to appease hunger – a loss which the 'rich farmers' can easily afford. For these occasional crimes the retribution brought down upon his head by the hunt is a wicked and cruel over-reaction.

Unfortunately and surprisingly, this caricature of the fox now seems to have been accepted by various 'experts' and even some BBC natural history programme makers appear to present the fox in the same way. Indeed the BBC's recent series 'Wild in Your Garden' could almost have been made by Walt Disney. Even the cover of the Corporation's listings magazine – *Radio Times* – showed the urban presenter with a fox cub.

The actual life of the fox is a fascinating one, and it is another strangely ironic twist that in the past, when foxhunting was in its hey-day, fox numbers were at their lowest and hunting was only a sport. Now that hunting is under threat, fox numbers are at their highest ever. Some form of control is required, and hunting would seem to be the most humane way of applying it.

The increase in fox numbers has been so great, over recent years, that cities have been

A fox cub – will grow up to be less cuddly.

colonised too – giving rise to the phenomenon of the much publicised 'urban fox'. The arrival of the urban fox has again worked against foxhunting, for in the cities foxes can do little real damage; they can create havoc with wildfowl collections in city parks and they may steal the odd pet, but in the main they live by scavenging. Dustbins, rubbish-tips and bird tables all provide good snacks; while some individual foxes will become so regular in their visits that milk and household scraps can be left out for them. In addition, cemeteries, playing fields and municipal parks and gardens can all make good hunting grounds providing insects, worms and small mammals.

Numbers have become so high in some urban areas that they are both caught and killed and trapped and released into the countryside all surreptitiously. If the situation was reversed and country foxes were trapped and released in towns there would be an outcry. But urban foxes are simply caught and tipped out into the country, regardless of what damage they might do. Ironically many of these dumped animals seem quite unable to cope with their new surroundings and inevitably get shot. But of course out of sight, out of mind – catching and dumping conveniently overcomes an urban problem. Yet to the conservationally correct; country people are not supposed to have a fox problem. The reality is that yes, country foxes also scavenge for their food, but they also include hens, lambs, pheasants and even avocets in their diet; temptations that rarely cross the urban fox's path – or pavement.

The fox population explosion has been underway for about forty-five years. In my childhood foxes were so scarce that I did not see one until my early teens. Now, if I see two during a Sunday afternoon walk around the farm fields I am not surprised. The change in the fox's fortune came with the introduction of myxomatosis among rabbits. Earlier it had been assumed that the rabbit appeared regularly on the fox's menu, and it was widely predicted that with this item of food gone, the fox would be very hard pressed to survive. In fact quite the reverse happened. When the rabbit disappeared, so did the rabbit trappers; they were responsible for the deaths of thousands of foxes every year. If a fox began to visit a snare line, before the trapper, then the trapper's livelihood was affected and the fox had to go. As a result when rabbit numbers declined to such a level that trapping became a waste of time, a tremendous pressure was taken off the fox, and because it is such an opportunist there were plenty of alternative delicacies available.

Two other developments took place at roughly the same time – the gin trap was rightly outlawed (it was one of the simplest and most efficient of all traps, but also one of the cruellest), and the fashion of keeping backyard hens began to die out. In fact keeping hens at the bottom of the garden was not a fashion but a practical necessity. For many ordinary country families it was the simplest and cheapest way of obtaining eggs and the hens' requirements could be largely met from household scraps, garden rubbish and gleanings from the cornfields. The presence of foxes would mean a loss of both hens and eggs and so any raid would result in the guilty fox quickly being trapped, snared or poisoned.

A town fox raiding a bird table. Where hunting is impracticable foxes can be a particular problem.

With the advent of battery eggs and supermarkets the price of eggs fell at a time when people were becoming more affluent and their work patterns were changing – so, it became cheaper and less bother to buy eggs. As a direct result many backyard hens disappeared, together with the snares in the hedge. This meant that the three greatest pressures were removed from the fox almost simultaneously and fox numbers have been increasing ever since. In addition to this, due to the number of animals and birds killed on our roads, an estimated 250 million a year, the fox always has a source of high protein food on hand and as a result no longer experiences winter food shortages.

Foxes will eat virtually anything from worms to geese, including voles, beetles, blackberries, plums, hens, fungi and pheasants. I even had a pet fox who would catch bumble bees in mid-air and eat them with relish. I have had several tame foxes and they would kill anything that wandered accidentally into their run, including birds, mice and on one occasion a hedgehog. Whenever they got out there was instant trouble, as they would make straight for the hens. The taking of an occasional hen would not cause a problem, but unfortunately foxes will engage in what is known as 'surplus killing'. If they get into a hen house they can leave many dead hens, while only taking one away to eat. Our biggest loss was 57 pullets (young hens on the point of laying eggs) in two nights, which on a small farm cannot be tolerated. The hens were not left out at night, but the fox broke into what was thought to be a secure deep-litter shed. In much the same way surplus killing can be a problem along hedgerows where pheasants are nesting or in a tern colony on a nature reserve. The 'experts' claim that when foxes kill more than they need they intend to return to the scene in order to make a cache of food for future use. In addition they say that in the artificial surroundings of a hen house the fox panics when feathers fly and seizes anything close at hand. I have seen foxes' surplus killing and believe that the true explanation is far simpler. My pet foxes were always very playful, a trait that went with them into adult life. This playfulness can be seen in wild adult foxes too. The hens I saw being killed early one summer morning were undoubtedly objects of play – with the fluttering feathers and hens in disarray being thoroughly enjoyed, just like dogs with a piece of wood – or out-of-control dogs chasing sheep.

It is because of the fox's wide diet and its habit of surplus killing that some control of fox numbers is necessary. Furthermore, because the variety of its food is so obvious to anybody who spends time in the country, those who attack foxhunting with the fashionable argument that foxes are harmless, innocent animals, are either showing great ignorance, or trying to mislead. Similarly those who claim that foxes cause no serious problems to wildlife are living in cloud-cuckoo land. One of the birds in most serious decline is the lapwing; on one reserve (sadly not an RSPB reserve) in the spring of 2003 a fox was shot. A post mortem examination revealed 32 lapwing chicks in its stomach. Chris Knights, farmer and winner of The Bird Photographer of the Year Award for 2003 has possibly the highest density of breeding stone curlews on his farm in Britain. The

Death in the hen house – bad for the farmer and for the hen.

stone curlew is one of our rarest ground nesting birds and without fox control Chris believes that the birds would not manage one successful nest.

The BBC must share some of the blame for creating foxes in cloud-cuckoo land as it regularly recycles all the old myths; foxes live on worms and voles; they do no damage; they do not kill lambs, and finally, of course, hunting is cruel. Amazingly, no mention is usually made of the fox's habit of surplus killing, nor the fact that both the RSPB and English Nature control foxes on some occasions when ground-nesting birds (not worms) are threatened.

In the course of writing this book I have come across several farmers who experience trouble with foxes after lambs, and one who has seen a fox stalking a lamb – he naturally intervened; but why is this simple journalistic exercise beyond those involved with BBC film-making? Before being closed down because of financial cuts, the old Ministry of Agriculture's experimental farm at Liscombe in Somerset showed lamb losses to foxes at 1% a year – out of total losses of between 5% and 6%. The Burns Report stated: 'The best estimate seems to be that a low percentage (less than 2%) of otherwise viable lambs are killed by foxes in England and Wales'. Although sounding insignificant in percentage terms – in real figures at 2% it means an astonishing 320,000 lambs taken by foxes a year Just 1% would yield 160,000.

It is incredible, fox apologists object to up to 25,000 foxes being killed by an assortment of mounted and foot packs of foxhounds a year – but they seem totally unconcerned at the figure of 320,000 lambs.

It is a mystery why or how the modem view is gaining ground that predators do not predate. Or is it simply dishonesty? A few years ago I visited a golden eagle's eyrie in England. It was situated in a beautiful, wide valley, full of ewes and their new lambs. It was obvious to me that the birds must be taking live lambs. When questioned one of the temporary wildlife wardens confirmed this. 'What do you tell RSPB members who come up here?' I asked. 'Oh,' he said, 'we tell them that the young eagles are being fed on dead lambs – they are dead by the time they get to the nest.'

It seems that, these days, lambs are also 'dead lambs' by the time a fox gets them to its earth. Earlier countrymen were not so blinkered, hence the old proverb:

> The fox barks not when he would steal the lamb.

Thomas Tusser (1524?–1580) also knew his facts:

> Keepe sheepe from dog,
> Keepe lambes from hog,
> If foxes mowse (bite) them,
> Then watch or howse them.

From his poetry and prose John Clare was a most accurate recorder of the countryside, and understood what took place in it. In 'The Shepherd's Calendar' he mentions:

> The sneaking foxes from his thefts to fright
> That often seize the young lambs at night.

Another thing that always puzzles me is why it is generally accepted that a fox will take a hare, which is heavier and faster than a newly-born lamb, yet people from the 'foxes do no damage to farmers' school of thought maintain that they will not take a lamb? Evidence from the Cairngorms suggests that foxes can take considerably larger prey than even hares. Alan Smith, with his wife Tilly, are Britain's only reindeer herders. One morning Alan was checking his herd at five o'clock when he arrived to see a fox making off with the head of a young calf in its mouth. The carcass was still warm; it was the best female calf of the year, twenty-one days old. Headless, it weighed 24 lb.

For those with an intimate knowledge of foxes the discussion ought to be concerned with the method of controlling foxes, not whether foxes should be controlled at all. Then, if an honest and accurate assessment is undertaken, the conclusion would be that the cruellest, but most efficient methods are snaring, shooting, and trapping; gassing was once used, but is now illegal. The most dangerous to other forms of wildlife are snaring and poisoning (which is also illegal, but still widely practised). The most inefficient, but least cruel method is hunting with hounds. It should also be remembered that it is often just one fox, or a small number, that is causing the trouble.

We have wild foxes visit the farm during most nights of the week, and on the whole we get few problems – but suddenly, we will get a spate of break-ins, accompanied by missing hens and chaos. It is as if an individual fox develops both the taste and the knack for killing hens. Once that fox is removed, the trouble stops. We also have a few guinea fowl; when a fox breaks in, the guinea fowl will always be the first to go – again suggesting that it is a matter of taste. Some gamekeepers, farmers, shepherds and wildlife wardens have all noticed the same thing – that fox damage is a matter of both opportunity and individual taste.

Most wildlife bodies with reserves in the countryside control foxes when certain species (usually ground nesting birds) are threatened, the RSPB, English Nature and various county wildlife trusts all control foxes and some other predators, usually to safeguard nesting wildfowl, waders and terns, but few will openly talk about it and some prefer to lose their wildlife than control their foxes. If no action is taken, then problems result. In my view the RSPB was wrong to have taken no action in 1983 to protect the nest of a pair of black-winged stilts in Cambridgeshire. It was the first breeding record of these most attractive birds for nearly 40 years. Despite a round-the-clock watch being kept on the nest to prevent disturbance by egg-thieves and 'twitchers' (obsessive bird-spotters who will travel miles to see a rare bird), the eggs were taken by foxes. As the RSPB reported at the time: 'Foxes were almost certainly the culprits. They were seen in the area on several nights and fox prints were found all around the nest. The birds left the area immediately and have not been seen since.' In view of the very high population of foxes and the scarcity of black-winged stilts in Britain, it seems to have been an act of great folly not to have surrounded the site with snares or an electric fence or, better still, both.* Since then there has been long list of legally protected birds that conservation bodies have not protected from foxes and as a result both individual birds and whole colonies have been lost. These include little, sandwich, common and the rare roseate terns – avocets, blackgrouse, capercaille, curlew, lapwing etc; some of the losses can only be described as a conservation scandal.

The methods used by conservation bodies have been shooting, snaring and before its banning, gassing – the tools of the gamekeeper. Indeed the main difference between the wildlife warden and the gamekeeper is quite simply that the wildlife warden is protecting specific wildife, while the gamekeeper is protecting wildgame. Shooting and snaring are efficient, but can cause the fox much suffering. Shooting with a shotgun leaves many wounded victims – who experience slow and painful deaths as confirmed in a recent report for the Middle Way group of MPs (2003). Some 'antis' claim that 'expert marksmen' should be used – they seem to forget that many shooting men are expert marksmen, and that whenever there is a moving target there is no guarantee of a clean kill. During the horrors of foot and mouth in 2001 DEFRA often used 'expert marksmen' to kill cattle in the fields. Some of the results were disgraceful with badly wounded animals running wild and some 'experts' taking seven or eight bullets to kill an animal. The suffering and cruelty was outrageous, as was the silence from those MPs, animal lobbyists and media luvvies who become apoplectic at the mention of hunting. Equally scandalous at the time was the silence on the government's preference for bullets rather than vaccination.

For years I argued against the use of gas**, believing it to be cruel, after hearing numerous stories of animals surviving the attempted gassing, but being left in a very distressed condition. Despite this, over many years, government departments, conser-

*A pair of black-winged stilts nested successfully in Norfolk in 1987.
***The Hunter and the Hunted* – 1977

vation bodies, and even the RSPCA claimed that the use of gas was humane. The policy seemed to be – if you can't see a death, then it is acceptable, however gruesome the reality. Fortunately, following the Zuckerman Report of 1980 into badgers and TB, it was finally admitted that gassing was cruel and the gassing of badgers and foxes was stopped.

The cruelty of snaring does not need describing, but it is significant that over 50% of the foxes killed by my local hunt – the Cambridgeshire – have snare or shotgun injuries. What will happen to these unfortunate animals if hunting is stopped?

Another alternative method of control is live-trapping – in which a fox is lured or deceived into a wire cage; that too is often erroneously described as 'humane'. It is not, for a fox held in a confined space feels great fear, particularly when approached. Sometimes it will work itself into a frenzy, damaging its claws and teeth in its desire to gain freedom – this too is one of the cruelties of snaring. A wild animal that is free has a number of options to avoid enemies or pursuers and feels no great fear. It simply tries to put what the scientist calls an adequate 'flight distance' between itself and what it regards

'That beautiful word, Fox, gladdens my 'eart' – Jorrocks.

as danger. If it cannot maintain a sufficient distance, it is caught and killed, but if its 'flight' is successful, normality quickly returns. On seeing a human presence in a field a fox will initially run, but on creating a sufficient 'flight distance' allowing it to feel safe, it will often stop and observe the scene at its leisure. Trapped animals have no way of creating a 'flight distance', and as a result experience terror. So, just as I objected to 'humane gassing' as cruel, I also consider 'humane trapping' to be inhumane.

The informed view can only be that hunting is the most humane form of fox control. Those that escape, by creating an adequate flight distance, quickly revert to the normal patterns of their lives, and those that are killed, die extremely quickly – there is no such thing as a wounded fox or a lingering death after a fox hunt. Recently I met a landscape gardener who had been working when a hunted fox was caught only 20 yards from him. He said: 'It was amazing. It was over so quickly that I did not realise at first that the fox had been caught. There was this great melee and it was finished in seconds. It must have died almost instantly.'

By its very nature hunting accounts for many old and sick foxes, which are often the ones causing the problems to farmers and shepherds anyway. Hunting also has one other benefit; as the hounds and huntsmen pass through they 'stir up the foxes'. It pushes them further out into open country, away from hen-houses and lambing pens, and makes them more wary of people and barking dogs, which in itself reduces the amount of fox damage.

The best and most sensible justification for foxhunting was put by a northern biology lecturer:

> 'The fact that fox damage can be serious, even with control as at present, can be confirmed by reports from various organisations which run reserves – RSPB, English Nature and local Wildlife Trusts, etc. Of the various methods of control only hunting is biologically acceptable as it is as near to the natural process as possible; it cannot wound, doesn't kill or injure other species (as do traps, snares, poison, etc.) and tends to select out the old, sick, injured animals which are the very ones most likely to do damage.'

Hunting limits the fox population but there is no danger of it being wiped out, even locally. Few countrymen would want to see the fox exterminated altogether, but many think that there are currently too many.*

The fact that huntsmen actually preserve foxes, sometimes creating artificial earths

---

*Unfortunately there are now many places, such as near towns and motorways and on some moorland, where there are large numbers of foxes and hunting is either impractical or simply cannot control them adequately. Here other methods have to be used and of them shooting with a rifle at night is best, if it can be done safely.

to encourage them to breed, is not a contradiction either. It again safeguards the long term future of the fox, particularly in areas of intensive agriculture. It also emphasizes how huntsmen are not the cruel sadists they are made out to be, as most of them actually love the fox. Again Jorrocks sums up their feelings best:

> ' Oh, how that beautiful word, Fox, gladdens my 'eart, and warms the declinin' embers of my age. The 'oss and the 'ound were made for each other, and nature threw in the Fox as a connectin' link between the two. He's perfect symmetry, and my affection for him is a perfect paradox. In the summer I loves him with all the hardour of affection; not an 'air of his beautiful 'head would I hurt; the sight of him is more glorious nor the Lord Mayor's show! But when the hautumn comes – when the brownin' copse and cracklin' stubble proclaim the farmer's fears are past, then dash my vig, 'ow I glories in pursuin' of him to destruction, and holdin' him above the bayin' pack!
>
> 'And yet (added Mr. Jorrocks thoughtfully) it arn't that I loves the fox less, but that I loves the 'ound more'.

It is hunting's links with conservation and the preservation of the countryside that are its most important aspects. This can be seen clearly in the area hunted by the Fitzwilliam foxhounds in parts of Cambridgeshire, the old Huntingdonshire and Northamptonshire, including the parish of Helpston, around which John Clare wrote much of his poetry. Clare would often visit Milton Hall near Peterborough, the family seat of the Fitzwilliams, where Earl Fitzwilliam was one of his benefactors. The Fitzwilliam hounds probably date back to the reign of Richard II, and are among the four oldest packs in Britain. The hounds still meet at Milton Hall, as they do at Helpston.

Helpston is not simply a reminder of John Clare, it is a living, evolving village; so much so that if the poet had been alive today his madness would probably have struck him down much earlier. In addition to its old cottages, Helpston has ribbon development each side of a long straight road into the Fens; box-like bungalows and dreary semi-detached houses.

Beyond Clare's village, the mixture of old and new continues; the black fen soil is divided into large hedgeless fields for sugarbeet, carrots and wheat. Where the soil changes and limestone creates a gently undulating landscape, large fields persist, some cultivated right up to the edge of the road, with ditches flail-mowed for tidiness. These are two of the hallmarks of modern 'efficient farming', but they are disastrous for wildlife. Some of the ugliness around Helpston is made even worse by electricity pylons, erected with little thought for the landscape.

Fortunately there are other farms, where the hunt is welcomed, with roadside hedges, spinneys and woods, many of which would have been walked by Clare. On these farms the fox and the pheasant live well and it is realised that there is more to farming than 'cash flow' and the application of sprays. Close by, too, there are the Barnack holes and hills – fifty acres that have been turned into a nature reserve. It is an area where the old limestone wildflowers still grow, including the pasque flower and several orchids

ABOVE: *Where meadowland flowers still grow.*
OPPOSITE: *'Tired but Victorious' by John King.*

– the 'bee', 'fragrant', 'pyramidal', and the extraordinary 'man'. Foxes, too, enjoy its cover, as well as its many holes and hollows. It is an area that was quarried for building stone – fine quality limestone, which was used by craftsmen to build the cathedrals of Ely and Peterborough, several Cambridge colleges as well as the ordinary cottages of the area.*

One of the most attractive stone villages is Titchmarsh, on the eastern edge of Northamptonshire. I visited it on a hot day in late July; it was almost like travelling into the era of Clare. It was quiet, apart from swifts, swallows and housemartins flying overhead, and there was a calm that, even on a warm day of high summer, I have never been able to experience in a town. The cottages were nearly all of stone, and many of the gardens were full of runner beans, marrows and potatoes, with clumps of hollyhocks by the back door. They were how country gardens used to be before the advent of imported suburban tastes, weedless lawns, flowering borders and garden gnomes.

I stopped at a large farmhouse, again made of stone, to meet Joan Wood. She is a working farmer, with 900 acres, some of which she rents and the rest she owns. In addition to farming she also loves hunting, and her land is hunted every year. Before showing me the extent of her farm, she had to check her calves, as she had just received some new ones which were 'blaring'; they had not yet managed the technological cow, with three teats, which kept its powdered milk at real cow temperature. It was good to see her working with the young animals, for there is always a close bond between a good stockman – or woman – and the animals. She also has 250 ewes; she is her own shepherd and does the lambing herself. Because she is a working farmer she has no time for gardening in her large walled garden, and so employs a pensioner, part-time, as a gardener. To make his work easier water has been laid on, to enable him to use a sprinkler; he is not impressed and continues to use his watering-can.

Her farm clearly shows the influence of hunting, made clearer to me by my drive to meet her. It took me through part of Cambridgeshire and the middle of old Huntingdonshire, which contains some of the worst treeless 'prairies' in the country, growing almost entirely corn, with one or two blocks of rape. Yet her farm was attractive and divided almost equally between grass and cereals; she loves livestock, she rotates her crops, she has kept all her hedges and her fields are no more than 30 acres in size. Her hedges are important to her for three reasons: they keep in the stock, they are good for hunting (hunt jumps have been put into some of them) and, most importantly, she likes them. The strange love affair between those who hunt and the foxes they pursue could also be seen, for she makes deliberate holes in her sheep netting so that foxes can move from field to field more easily. She never has any fox trouble with her

*'The Hills and Hollows' National Nature Reserve

lambs; she believes that this is because she lambs indoors. Then, by the time the lambs meet their first fox they are strong and mobile. She also believes that a ewe will see off a fox. This is certainly true, for sometimes I take my little lurcher, Bramble – a cross between a whippet and a Bedlington terrier – through a small field of sheep. He is about fox size and when lambs are present the ewes will advance, stamping their front feet and making noises of agitation – he keeps well away from them.

Each year Joan Wood grows a field of kale, both as fodder and for the benefit of wild foxes. It is a mystery why foxes like kale, as the leaves are notorious for holding water and they always seem to be cold, damp places in winter; yet almost invariably when the hunt visits her land a fox is found in that field. Perhaps it is the strong smell of kale which the foxes like, the shelter from the wind, or the total concealment from passing eyes.

On a field of permanent pasture the old ridges and furrows showed the cultivations of an earlier age. Rooks were calling and flying above a rookery in a small oak spinney; she has several ideal sites for rookeries on her land. Both rooks and jackdaws were flying and tumbling in the thermals. They were good to see, for they are two more casualties of intensive farming. Jackdaws have disappeared almost completely from my parish, with only occasional small groups passing by in winter. There are simply no old trees, with holes, in which they can nest. It is an alarming and saddening change; in my childhood they were everywhere, nesting in the many pollarded elms and their calls were an everyday part of spring and summer. They made good pets, too, usually called 'Jack'. There were stranger nests in the old elms as well, for I found my first pet fox cub in a hole in an old elm at least eight feet from the ground.

A quiet spot of summer – preserved for the hunt jump in the winter season.

There are several spinneys on Joan Wood's farm; fox coverts to provide both cover for foxes and an attractive landscape. Where Dutch Elm has been active she has replanted with indigenous species, including oak and ash. At the edge of one newly planted spinney, by a field full of big bales, I noticed a heap of small bales, specifically placed to provide fox shelter during the winter. Around an old pond the temptation to 'fill in' had been resisted and she had planted more trees, fencing them off to avoid damage from stock, rabbits and hares. As I walked over to look at the young trees I

A flail-mown hedge in summer – bad for the fox and all other forms of wildlife.

flushed two hares; they do very well on the farm, thriving on both her mixed farming and her sympathetic methods.

Titchmarsh meadow is one of the most interesting places on her farm. It is an ancient wet meadow and a Site of Special Scientific Interest. The fact that it is an SSSI does not worry her and she has no plans to drain it, spray it or change it in any way. Some farmers, who do not know any better, dislike having SSSIs on their land and distrust English Nature who designate them; part of the problem is caused by the title itself, as 'SSSI' is rather intimidating. Perhaps the areas should be called Sites of Special Wildlife Importance – their function would then be immediately obvious.

Joan Wood's meadow is small, being only 5.6 acres in size, but it is very special.

The land lies by a stream, it is wet and wild, surrounded by overgrown hedges and trees. At one time it contained a medieval fish pond, and later rushes were harvested for thatching. Now it is a paradise of wildflowers, ant hills, and insects that give a confusion of wildlife riches. Botanically it is very interesting with a variety of grasses, rushes and sedges, including quaking grass, crested dog's tail, blunt-flowered rush, reed-canary grass and, appropriately, false-fox sedge. On the day of my visit ragged robin, creeping jenny and greater bird's foot trefoil were all on flower; I was just too late for the mass of flowering southern-marsh orchids – a Northamptonshire rarity. The field was full of grasshoppers and various crawling, hopping and flying insects, as well as damsel flies. Again it gave me a glimpse back into my own childhood, for grasshoppers, too, have been hit by the use of sprays, fertilisers and flail mowers. A partridge and her brood of newly hatched chicks skulked, and numerous small holes and paths in the tangled vegetation showed a high population of voles. The land is not left entirely as a wildlife wilderness, for each year in late summer it is grazed, which benefits and stimulates the plant life. For foxes it gives cover and good hunting; for birds it provides nesting sites and food, and for butterflies and grasshoppers it is an ideal place in which to breed. To Joan Wood, it adds to the attractiveness of her farm; it would represent a loss and a needless waste of land to the 'agro-businessman'.

In much of Britain the hunting landscape is patchy, and often separated by great tracts of factory farmland. But in the Midland shires, where foxhunting first became fashionable, it still dominates much of the country. To see that also at first hand I visited the tiny village of Gumley in Leicestershire, the home of Lt.-Col Murray-Smith, a landowner, and one time master of the Quorn and Fernie hounds.

He had hunted since he was very young, but it had provided him with more than simple sporting pleasure, for his land created an attractive rolling landscape – a patchwork of grass, corn, hedgerows and woods – many planted in the first half of the nineteenth century specifically for hunting. The imprint of post-enclosure hedge-planting is still clear to see, and the hunt jumps cut into them show why many have been retained.

I was shown the area by a remarkable old countrywoman and her Staffordshire bull-terrier. It is hard to classify 'Pop' Payne. She is a mixture of gamekeeper, estate worker, earth-stopper and naturalist, whose knowledge and understanding of the wildflowers and creatures around her have come from a lifetime of living and working on the land, not from textbooks and political manifestoes.

The woods and shelter-belts are of oak, ash and sweet chestnut – all the elm of the area died several years ago. The hedgerows also contain many mature trees at intervals of about fifty yards; they were planted to provide posts and rails for fencing – the oak for posts and the ash for rails. In addition both trees provide good habitat for wild creatures; foliage for cover, nesting sites, acorns, ash seeds, and ideal conditions for a multitude of insects.

In the spinneys and coverts, the wealth of birdlife can be heard in the songs of warblers and chiff-chaffs, as well as the laughing cry of the green woodpecker. Not only do the woodpeckers have plenty of holes for nests and bark in which to feed, but some of the grassland also contains ant hills, which provide a valuable source of food.

Scores of foxgloves and red campion grow among the trees, as well as large bellflowers and wild raspberries. Early in the year some of the woods have bluebells, wood anemones and primroses – they have to be protected – not from ploughing or farm sprays, but from people who illegally pick them, or dig up whole plants for their gardens. Where elms have died the woodland has not been lost, for Pop has replanted with oak, beech and walnut. Wildflowers are plentiful along the hedgerows, and they are often helped by the fact that around several fields 'hunting headlands' are left. These are uncultivated and unsprayed strips of land which allow horses and riders to pass from field to field during the winter hunting season, without damaging crops. Along one hunting headland we saw a large grass snake; a creature scarce on most modern farmland. Willowherb, both the great hairy and the rosebay, is particularly common. The rosebay is the most attractive and at one time it was encouraged into old country gardens to attract bees.

One unsprayed grassy bank was like a wild garden with field scabious, knapweed and lady's bedstraw in full bloom. Butterflies were numerous – large whites, skippers and meadow browns, as well as a daytime-flying moth, the six spot burnet. Pop's name for the meadow browns was most appropriate, she called them 'mowing grass butterflies', as they appear at about hay time. She used several local names; goosegrass she called 'herrief' and yellow rattle was 'rattle grass' or 'wattle'. She spoke of 'totter grass' too but she has not seen it for several years; it was probably quaking grass which has disappeared over much of Britain. Because of the richness and variety of the land, hares and English partridges do well, and despite the presence of foxes there is enough game to allow rough shooting.

Badgers also thrive on the estate, with numerous setts, some in the woods and along the hedgerows. One new hole had simply been dug out in the middle of a field of barley.

Outside an old sett entrance was the skull of a lamb, but Pop was not worried: 'The badgers do not kill lambs, but they will eat any they find dead. We do not get much fox trouble with lambs either.'

As we walked back at the end of the day, Pop found the first harebells of the year, a find that gave her much pleasure. We returned over another part of traditional England – the village cricket pitch. The 'square' was the only part rolled and cut, with the bumpy outfield mown by sheep, and divided in two by a small country road. Pop enjoys her earth-stopping: 'It means I'm out a lot at night, while those in towns are complaining about the lack of light or sitting inside worrying about being mugged. I love foxes – I often see them.

'I enjoy watching the hunt too; it will be a shame if they stop it; it's part of England.

'Away from Clawson Thorns' by John King. This is a famous view of the Belvoir Vale.

Those who complain about it don't understand it. The young antis are just being used as political fodder.'

In Cambridgeshire, Northamptonshire and Leicestershire I had visited three areas of hunting country*, virtually at random. All had been attractive, with hedgerows, birdsong, butterflies and wildflowers. I had seen the sort of countryside farmers are always being urged to retain, or create. Yet without the presence of the fox, the main features would have been very different – acres of winter wheat, flail-mown drain edges and wildlife silence.

The links between hunting conservation, landscape and wildlife are obvious to anybody with an open mind. As the Burns Report confirms: 'Foxhunting has undoubtedly had a beneficial influence on lowland parts of England in conserving and promoting habitat which has helped biodiversity, although any effect has been in specific localities.... Hunting has clearly played a very significant role in the past in the formation of the rural landscape and in the creation and management of areas of nature conservation.'

Chapter 3

The Hare in the Meadow

It is a measure of the amount of damage done to the British countryside over recent years that hares should be included in this book at all. A few years ago the ordinary wild brown hare was so common that it was taken for granted. On almost any country walk several hares would be flushed and the sight of a hare disappearing into the distance with a barking farm dog floundering along behind made an amusing break for the farm-hand working in the fields. Even the hunting of hares was largely ignored, although there has always been a body of opinion opposed to hare-coursing; a sport in which the speed of the hare is pitted against that of the 'long dog' – the greyhound, whippet or lurcher.

Over recent years things have changed and hare numbers have plummetted in many areas. Strangely, until recently, most conservation bodies and anti-hunting organisations seemed unconcerned at this demise and I know of no reserve that has been set up specifically to preserve or to monitor the status of the hare. Those who hunt or course, however, became alarmed early on and took special measures to preserve the hare; it is an example of the hunter ensuring the welfare of his quarry.

The Game Conservancy Trust is a charity whose membership is made up largely of land owners and those interested in country sports, particularly shooting. Its function is to undertake proper scientific research into the problems faced by game and wildlife, and it is the one organisation to have taken the decline of the hare seriously, researching into its behaviour and the hazards now confronting it. The findings predictably confirm that the hare is under increasing pressure and that numbers began to fall in the early 'sixties.

In my part of East Anglia the decline of the hare was both sudden and tragic. On our farm we did not see a hare in its March madness throughout the 'eighties. It was a sudden and tragic population crash.

It was a sad loss, as the hare has been an interesting part of country life and lore for hundreds of years. At one time, too, it was also a welcome and important addition to the countryman's dinner table; even now 'jugged hare' with stuffing balls and red-currant jelly is a meal that cannot be bettered, and hare gravy makes a mockery of those thin brown liquids called 'gravy' on television advertisements.

In our family the gravy is important, for whenever we eat game or poultry we start the meal with 'light pudding' – as we do for our traditional Christmas dinner. It is like a savoury sponge and has to be covered with spoonfuls of gravy, for it is also known as 'blotting paper pudding' and is delicious. The ingredients are simple – 8 ounces of self-raising flour; 1 teaspoonful of baking powder; 3 ounces of margarine, two eggs, and a small quantity of milk. The margarine is rubbed into the flour and then the beaten eggs stirred in. Milk should be added, but not too much; the mixture should be quite stiff and definitely not runny. It is then baked for twenty minutes, until it has risen and the top is crisp. In this quantity the pudding is large enough for between six and eight people. It is a recipe that has been handed down by word of mouth through several generations.

Those who taste it for the first time invariably take the recipe to use themselves, but despite searching I have never seen it written in a cookbook and have no idea of its origin.

Despite this, some old country people refuse to eat hare, even where it still thrives, for it seems to have a greater volume of blood in its body than other animals of its size, which makes the meat dark. As a result it is believed that if a pregnant woman eats hare she will have a miscarriage. The meat of a 'bucking hare' is also avoided as it is said to cause stomach upsets.

Because of their wide wild eyes another country belief once suggested that witches turned into hares. If a circle of hares was seen, with all the hares looking inwards, then it was thought to be a witches' coven. The old link between witches and hares is shown in a short poem by Walter de la Mare.

> In the black furrow of a field
> I saw an old witch-hare this night;
> And she cocked a lissome ear,
> And she eyed the moon so bright,
> And she nibbled of the green;
> And I whispered 'Whsst! witch-hare,'
> Away like a ghostie o'er the field
> She fled, and left the moonlight there.

Another peculiar belief in some areas, involving bad luck, claims that if a hare runs through a village street, a fire will break out shortly afterwards. To possess a hare or

rabbit's foot, however, is thought to bring good luck and cure rheumatism. There were children attending my local school 50 years ago who still carried their lucky foot.

Hares deserve some of the peculiar country lore associated with them, for they are very highly strung. This makes them extremely difficult to keep and breed in captivity, as they frighten easily and can damage themselves by running into fences. Several years ago when a neighbour found a leveret, she thought she was succeeding, until the unfortunate animal chewed through the television's electric cable and was electrocuted.

The poet William Cowper met with more success and kept a hare for over eight years, leading to one of his most famous poems – 'Epitaph on a Hare':

> Here lies, whom hound did ne'er pursue,
> Nor swifter greyhound follow,
> Whose foot ne'er tainted morning dew,
> Nor ear heard huntsman's hallo'. . . .

His love for his tame hares also helps to explain his aversion for hunting.

Much of the hare's natural history is still a mystery, including its March madness; for although it breeds throughout the summer, its most erratic, or erotic, extravagances are to be seen during the month of March. Nearly all leverets are found singly, yet some countrymen believe that the mother gives birth to her family in distant parts of a large field to avoid losses from foxes, while others say that all the leverets are born in the same place and then moved to different 'forms' (depressions in the ground) by the doe (female). Many gamekeepers also claim that the mother hare visits each of her litter in turn to suckle them, although others maintain that she calls them up and feeds them together.

Another peculiar observation, repeated in some natural history books, is that if a leveret is found with a white spot on its forehead, then there are three in the litter. Old Jim, on our farm, believed that story. He also stated that when resting in a form, a hare spends most of its time looking backwards, consequently foxes are said to approach from the front. He always called a hare 'Sally' (another old country name is that used by Cowper - 'Puss'), and claimed that 'winter wheat should cover a hare in March', and that 'ten hares eat as much as a sheep'. His foolproof method of telling a hare from a clod of earth was simple: 'As you walk towards a piece of dirt it seems to get bigger; when you walk towards a hare it gets smaller, skulking into the ground.'

But it is the hare's speed that has always fascinated people, and for centuries a running hare was regarded as a more noble quarry than the fox. Queen Elizabeth I had her own pack of beagles and coursing, with 'gaze hounds' dates back to the time of the Ancient Egyptians. Even now those who work in fields where hares are plentiful get satisfaction, and entertainment, from seeing their dogs chase a hare, and if a dog gets close enough to make the hare 'turn' (suddenly change direction) then it is a fast dog.

A sough – an artificial refuge created to give the coursed hare shelter. The hare's normal shelter ('form') is just a depression in the ground.

We once had a young labrador who managed this, but the change of direction was always so sudden that it would leave the dog sprawling as the hare made for the safety of the nearby hedge.

My little lurcher, Bramble, only saw one hare in the first ten years of his life, and that was disappearing into the distance at high speed. A lurcher is the traditional gypsy poaching dog – now fashionable with all types of people – a cross between a whippet, greyhound or even deerhound (for speed), and a labrador, collie or terrier (for intelligence). They are supposed to hunt and run silently. Most of Bramble's runs were after foxes; he then committed a most un-lurcher-like sin, he barked at them.

From the research undertaken by the Game Conservancy it is clear that there are several reasons for the hare's plight. Like many animals it thrives on a varied diet, and so huge areas of mono-culture wheat were not to its liking. In such an area food can be desperately short, particularly straight after harvest when the land is ploughed and the odd unsprayed green shoot is a rarity. This was once true of the land surrounding our farm, where, by the middle of September, the landscape was almost entirely of bare earth.

In other areas of grass farming for livestock, modern farming methods have created yet more dangers for the hare, and the forage harvester, used to cut and collect grass for silage, claims many thousands of small leverets every year. Sprays, of various kinds are another hazard. Gramoxone is used to 'burn off weeds', and wide spraying arms can also burn off hares, causing them a long and painful death. Instead of running at the approach of the sprayer, the hare crouches and gets sprayed; when it licks itself clean, it gets poisoned. Some farmers add to the problem by spraying in the evening; so when the hares emerge to feed, the poisons are still wet and active. In addition, several sugar beet and carrot sprays are highly toxic, killing not only hares but also pheasants, partridges and many song birds.

'Mono-germ' is another modern-day threat to the hare; at one time sugar beet was sown thickly, and the growing crop was thinned by hand. Then, hare-damage could be tolerated. Now sugar beet is sown with each seed correctly spaced out and guaranteed to germinate, so hares can cause a much greater loss. Consequently the farmer who is not interested in wildlife, seeing game, or having the beagles (hare hounds with foot followers) or harriers (hare hounds with horses and riders) over his land, will shoot all the hares he sees. Shooting grazing hares using a .22 rifle with telescopic sights, sometimes at night with the aid of a strong light, can virtually wipe out the entire local hare population. Landowners who indulge in this, or allow it to take place on their land, are usually the ones whose main interest in life is their own bank balance and little else.

Some landowners also eliminate hares to avoid the threat of illegal hare coursers. Illegal hare coursing is simply undertaken by people without permission to be on the land – consequently not only is it trespass, but it is also poaching, theft. Many didecoys and travellers are involved and if challenged can become violent – even firing steel ball bearings at gamekeepers with catapaults. They do nothing to aid hare conservation and unlike legal coursing, in which the aim of the course is simply to run the dogs against a hare – the winner in illegal coursing is the dog that kills the hare. Often they observe no close season and large amounts of money change hands. (Legal coursing has had its own close season between March 11 and September 14 since 1851) Ironically those who rant and rave about the evils of legal coursing – in which just 250 hares get killed a year – have little to say about illegal coursing; the police are largely uninterested and ineffectual and any law to make coursing illegal would not affect 'illegal coursing' – it is 'illegal' already and rapidly growing.

From the Game Conservancy's research it is clear that the presence of hares today is usually a sign of a comparatively healthy countryside. One in which there is a choice of crops; where spray is used with discretion, and where wooded shelter belts and hedgerows break up the land. Such countryside is often found where pheasant and partridge shooting takes place, and so the presence of hares is one of the bonuses of good game-bird management.

I saw this confirmed on one of the most famous shooting estates in Britain, situated on the eastern boundary of Cambridgeshire. The estate covers some 7000 acres including 2000 which are farmed 'in hand' by the owners. Included in the 2000 acres are 140 acres of sugar beet and 350 acres of woodland — a mixture of hard and soft woods, as well as grassland and cereal crops. Two small woods near a road, which harboured more poachers than game, have been removed; but fifty acres have recently been planted to compensate, with softwoods, beech and oak.

Hares have always been a feature of the estate and no hare is shot during a pheasant shoot. In late winter special hare shoots are held, however, to reduce numbers, for otherwise they would cause much damage; but even then the areas of woodland and shelter belts are not driven, to ensure that a good breeding stock is always preserved.

Some hares are occasionally netted, to re-stock areas where they have become scarce; this is only done for owners whose farming practices have been modified to ensure the new hares' survival. As a result, hares have been successfully re-introduced to several areas including Windsor Great Park and an estate in Herefordshire. In addition, new stock has also been sent to the coursing grounds at Altcar, in Lancashire, after a decline in the coursing interest led to a fall in the local hare numbers.

Keeping the hare population healthy has not been easy. Noel Cunningham-Reid, who runs the estate, believes that the greatest dangers facing the hare come from farm machinery, sprays, foxes and cold wet weather, particularly when the leverets are small. Consequently, on his land the use of Gramoxone is banned, and he is currently worried by the affects of a certain chemical fungicide, and is looking for a less toxic alternative.

Despite the care taken, the hare population has fallen by fifty per cent over recent years, although he hopes the slide has now been halted. About 700 are shot every year, whereas a few years ago the figure was well over a thousand. A new motorway has been particularly damaging along one side of the estate, bringing both a high number of road

Hares in their March madness.

casualties, and poachers. Along another boundary where the adjoining land has no gamekeeper, numbers are also low because of fox predation. Both foxes and stoats take many leverets on un-keepered land.

There is no organised coursing on the estate, but didecoys poaching with their lurchers are creating an ever-increasing problem. At times there have been over twenty men and their dogs, and some of the vehicles (the few that are licensed) have been traced from as far away as Kent. Yet the police are becoming increasingly reluctant to become involved, because of their own workload, and because of the number of trespassers and poachers involved. It has to be said, as well, that the modern police force is becoming more and more urban and suburban orientated, and shows little interest in rural problems. The old type of local country 'bobby' would have willingly helped round up poachers, but today's motorised policeman is most reluctant to leave the warmth of his patrol car, or exchange his shiny leather shoes for Wellington boots.

I was shown over the estate by the head gamekeeper on a bright November day, when the beech leaves were the colour of burnished copper. It was beautiful landscape, with woods, hedgerows and shelter belts breaking up the arable land, and pheasants feeding along the field edges. There were rooks and jackdaws, too, as well as skylarks. Several lesser black-backed gulls circled in the air, and one was feeding on the carcass of a rabbit. The keeper notices them every year: 'Each autumn they arrive at about this time and hang around for a few weeks.' If there had been no game interest, and the land management had been concerned only with maximum profit and production, it would have been one large featureless prairie. Instead it was 7000 acres of almost traditional landscape, with many areas of great wildlife value.

The 'keeper takes a special interest in the hares and talks of them with affection. 'I love watching them – especially in the spring when the bucks are boxing. They get in such a frenzy when half a dozen bucks are chasing one doe. If you stand still they will come so close to you that you can hear them panting and talking to each other.'

His interest goes beyond game, for he is also a good naturalist and is one of the growing number of gamekeepers who does not see all predators as a threat to his game and is pleased that buzzards have recently started to breed in the area.

'I respect birds of prey and enjoy seeing them. We've got breeding kestrels, tawny owls and little owls. They are no problem. We have the odd rogue tawny, but that can be tolerated. I've not seen a barn owl for four or five years – that is really sad. We've had a pair of hobbys breed here successfully for two years running; there is an odd hen harrier about most winters and we have a few sparrowhawks. It's funny really, we had a pair of Montagu's harriers breed a few years ago. The RSPB had to station a man here all the time to keep people away. Two years earlier we kept it a secret and had no trouble at all.

'I've got a problem at the moment. We have stone-curlews breeding regularly. This year we had three successful nests. I look out for them and let the tenants and the tractor drivers know. I want to let the RSPB know, too, but if I do we'll get flooded out with people doing their utmost to disturb the birds, simply to see them.'

To assist insect life, which is beneficial to both game and wild birds, the margins of the field are not sprayed. No flail mowers are used on the estate either, apart from a little tidying of the roadside hedges in late summer. The head-keeper hates flails: 'They are terrible. They are the great enemies of wildlife today. They damage hedges and trees; they are bad for insects and ant hills and they destroy a lot of suitable habitat for butterflies and moths.'

He took me to one area of rough grassland that each spring is carpetted with cowslips and has a summer-long succession of butterflies and grasshoppers. Nearby was a hedge planted to give shelter to pheasants; it was buckthorn, the foodplant of the brimstone butterly. Occasionally, during the course of a year, he will get regaled by people opposed to shooting: 'If they are countrymen and understand the country I will listen to them. But I won't listen to those idiots who come out from the town and who don't know a bull's hind foot from a dog's ear. I won't have them buggers tell me what to do. Man has always been a hunter and so I've got no time for them.'

To go around such an estate, where people are aware of wildlife, as well as the game, is quite reassuring, especially when compared with the usual attitudes of the over-intensive farmers. The 'keeper is such a pleasant man that he is a good advert for his profession too. Many of those involved in the anti-bloodsports movement, especially politicians, would-be politicians and self-publicists, accuse gamekeepers of being 'cruel', 'callous' and 'barbaric'. I found the head-gamekeeper to be sensitive and understanding. I asked him what he would choose to do if he was not a 'keeper. He

Female kestrel resting on a hedgerow fence. They breed on the Six Mile Bottom estate.

thought for a moment and then said: 'The only other thing I would like to do is play the cello. I love music, and apart from my wife and my job, good music is the main thing in my life.'

In addition to land managed for shooting, hares are also surviving better on land where they are hunted or coursed. Dr Stephen Tapper of The Game Conservancy confirms this by saying: 'Hare coursing can only take place where hares are numerous, so naturally coursing people do their utmost to ensure their abundance. Places where coursing takes place often have very high hare densities.' With this in mind I went up to Altcar, then owned by the late Lord Leverhulme. It is the home of the Waterloo Cup – the most famous coursing event in Britain, dating back to 1836.

The whole area is extremely flat, bearing a marked resemblance to the East Anglian fens. In Lancashire the old, wet flat lands are known as 'mosses', although technically they are areas of lowland acid peat-raised bogs. At one time these places of marsh and bog covered over 50,000 acres in Lancashire, but most have now been reclaimed for arable agriculture. Virtually the whole of the estate at Altcar has been reclaimed; a fact confirmed by the numerous drainage dykes and water pumps, as well as the river behind a solid earth bank. Now, most of the land is free-draining with the peat soil being fertile and highly productive. It is interesting to note that until the First World War, almost the whole of the Altcar estate was grass, and, according to a local newspaper, 'provided all the hay for the working equine population of Liverpool'.

I visited at hay time, with the other growing crops also looking well. There were small fields of barley, wheat, potatoes, carrots, cabbages and the occasional patch of grass. The two main coursing fields for the Waterloo Cup were entirely of grass, each sixty acres in size, and every year they are cut and baled for hay.

The drainage dykes themselves, despite being well maintained, had a number of attractive water plants growing in them, including flowering yellow iris (yellow flags), and, much to my surprise, water violets. The water violet is quite a scarce plant; its petals are the same subtle blend of pink and pale purple as the cuckoo flower. But strangely, in spite of its name, it is an aquatic member of the primrose family and quite unrelated to the violet. Along the banks were cinquefoil, marsh bedstraw, and clumps of meadowsweet, alive with insects. The Indian immigrant, Himalayan Balsam was also present, as was the native but ubiquitous foxglove. A few areas of rough grass and birch trees completed the patchwork of moss and meadow.

Yellowhammers were calling 'a little bit of bread and no cheese', and in the dykes mallards swam for cover with their growing broods. In some fields oystercatchers stood watching with suspicion, sometimes piping in warning and irritation. In Lancashire, like much of Northern England and Scotland, they often breed well away from the sea. During the winter the sound of oystercatchers and the song of the skylark give way to the call of geese, as several thousand pink-footed geese pass through on migration, with some staying on to spend the whole winter in the area, feeding on the farm fields.

In these conditions hares are now flourishing, and again their survival has been helped by the banning of Gramoxone and some sprays used for carrots. There are other indications that the local farmland is healthier than most, as English partridges still breed and the estate also has resident barn owls. I saw several hares during my visit; one running away from my approaching feet with its ears down, and two on a farm track who briefly reverted to their March madness to box, before running off in different directions. On a small grass meadow, recently cut for hay, there were at least five feeding contentedly. It gave me great pleasure seeing hares in the fields again; it was like travelling backwards in time, reminding me of earlier days when hares were an everyday part of country life.

Because of the changing status of the hare it seems to me that any action taken to preserve it must be good. Consequently coursing and hunting ought to be acceptable on conservation grounds alone. Where sporting interests are keeping hare populations high, it is because they are creating a habitat suitable for its survival, which means fewer sprays, a variety of crops, and habitats which benefit many forms of hard-pressed wildlife. But inevitably the moral question also has to be answered - for is hare coursing cruel? And is coursing the most 'bloodthirsty' of all the 'blood sports' as many opponents and abolitionists insist?

Hare hunting is something that I cannot get excited about. William Cowper obviously thought that it was extremely cruel and wrote:

Detested sport,
That owes its pleasures to another's pain.

However, Wildred Scawen Blunt (1840–1922), took a different view in his famous poem, 'The Old Squire':

I like the hunting of the hare
Better than that of the fox;
I like the joyous morning air
And the crowing of the cocks. . .

. . . I like the hunting of the hare;
New sports I hold in scorn.
I like to be as my fathers were,
In the days ere I was born.

A mountain hare in its winter coat.

The complete poem is seventeen verses long and clearly shows the hunter's love for the countryside, tradition, the hounds and his quarry.

William Cobbett (1763–1835) also greatly enjoyed 'hare hunting', as well as coursing. He liked them better than shooting, for 'the achievements are the property of the dogs'. His support for country sports is interesting, for he was no lover of the Establishment or the aristocracy, but he liked cant and hypocrisy even less. He was an outspoken countryman – a radical and a free-thinker, and when he met a fool – which, according to him, was often – he said so.

He did not regard hunting or coursing as cruel, and in his *'In Defence of Blood Sports',* he used, in his own forthright style, many arguments that still apply today:

> . . .Others, in their mitigated hostility to the sports of the field, say that it is wanton cruelty to shoot or hunt; and that we kill animals from the farmyard only because their flesh is necessary to our own existence. PROVE THAT. No: you cannot. If you could it is but the 'tyrant's plea'; but you cannot; for we know that men can and do live without animal food and if their labour be not of an exhausting kind, live well too, and longer than those who eat it. It comes to this then, that we kill hogs and oxen because we choose to kill them; and we kill game for precisely the same reason.

A hare dozing in the summer sun – a picture from 'Lepus The Brown Hare' by BB, which shows the 'patchwork' landscape hares enjoy.

> . . . A third class of objectors, seeing the weak position of the two former, and still resolved to eat flesh, take their stand upon this ground; that sportsmen send some game off wounded and leave them in a state of suffering. These gentlemen forget the operations performed on calves, pigs, lambs and sometimes on poultry . . . Only think of the separation of calves, pigs and lambs at an early age from their mothers! Go you sentimental eaters of veal, suckling pig and lamb and hear the mournful lowings, whinings and bleatings; observe the anxious listen, the wistful look and the dropping tear of the disconsolate dams; and then while you have the carcasses of their young ones under your teeth, cry out as soon as you can empty your mouths a little, against the cruelty of hunting and shooting.

I have witnessed hare hunting with both beagles (hounds now mostly followed on foot) and harriers (followed on horseback)*. Whereas I found fox hunting genuinely exciting – at first hare hunting was most uninspiring for me, although I did enjoy riding with the Aldenham Harriers. Those who participate enjoy seeing the hounds 'work' — when the pack finds and follows a line of scent. The end is simple; the hare is either killed, or it gets away. This ought not to shock, as the hare's whole evolution has been based on flight and the evasion of those birds and beasts that prey on it. Hares are fast and alert and the minority that die to hounds do so far more humanely than those crippled on motorways, badly shot or poisoned in farm fields – and so in all honesty I cannot call hare hunting cruel, although hare hunting is not for me.

Of all the 'blood sports', coursing is the most reviled. Yet the death of the hare – on the few occasions that it is caught – is virtually instantaneous. Opponents of coursing claim that hares are 'ripped apart while still alive' – a physical impossibility considering the size of the hare and the fact that only two greyhounds or whippets are involved. Most of the hares die as soon as they are caught, with a broken neck, and the bodies are taken by those who course to be eaten as 'jugged hare'. Only occasionally do both dogs seize the hare at once, which prevents an instant death. Some opponents also suggest that the hares are released in front of the greyhounds; again this is often willful misinformation, for coursed hares are completely wild and free. They are either driven to the coursing field by a line of beaters – similar to a pheasant shoot – or else they are 'walked up' – the coursing party walks over a field where hares are almost certain to be found, and two dogs are slipped from their leads when a hare is flushed.

The other great error in the minds of most critics is that the winner of a course is the dog that kills the hare. In fact it is the dog that follows most skilfully, and there have been several winners of the Waterloo Cup that have not killed a single hare in their series

---

*When writing *The Hunter and the Hunted* and *The Hunting Gene. For The Hunting Gene* I rode with the Aldenham Harriers.

Coursing is one of the oldest sports – but never practised now on horseback.

of eighteen courses. The hares are always allowed a start of about one hundred yards; as a result six out of seven coursed hares get away.* This has always been the case, and as long ago as AD 116 the Greek historian and philosopher Flavius Arrianus (he had Roman citizenship) wrote: 'The true sportsman does not take out his dogs to destroy the hares, but for the sake of the course and the contest between the dogs and the hares, and is glad if the hare escapes.' According to the Burns Report just 250 hares are killed a year through legal coursing whereas between 200,000–300,000 are shot annually for control, and crop protection; even more die each year as a result of modern farming methods. The hue and cry over coursing would therefore seem to be quite out of proportion – especially when compared to the other more gruesome hazards faced by the hare.

I can understand the appeal of coursing as a hare does provide a real challenge of speed and agility for a dog. It is also easy to see how coursing became an organised sport, particularly when it is remembered that before the ages of radio, television and easy transport, rural communities had to make their own entertainment. Even my great grandfather, at the turn of the century, went coursing every Boxing Day in the Fens.

At one time the Waterloo Cup was the most important sporting and gambling event

*Every Waterloo Cup winner since 1984, including the winner in 1987, did not kill a hare at the Altcar meeting.

in Britain. In 1874 it attracted a crowd of 80,000 and as late as 1952 the draw was on television – rather like the FA Cup draw today. When Master McGrath, a dog owned by Lord Lurgan, became the first greyhound to win the cup three times in 1871, Queen Victoria gave the dog and its trainer a Royal Audience at Windsor Castle – to reward a popular sporting hero.

Since then the fortunes of the Waterloo Cup have fluctuated and between 1976-1980 hares died in large numbers at Altcar due to carrot spray and Gramoxone. Coursing men saw the dangers, the sprays were removed, and now the hares are even fed during the winter to keep them in good condition; oats, apples and artichokes are all left out for them in the fields, as well as willow sticks, for they love eating the bark.

Over recent years, with a restored hare population, the Waterloo Cup has experienced a resurgence of interest. One of those responsible for it is a Newmarket racehorse trainer, Sir Mark Prescott, who has become fascinated by both coursing and the natural history of the hare. His interest in coursing started nearly thirty-five years ago completely by chance, when he stopped his car, to see why a crowd had gathered in a field; it was his first experience of coursing. He enjoys it now and justifies it by saying that: 'Coursing is in the overall interest of the hare.'

He is an articulate and amusing man who objects to the double standards adopted by society whenever the subject of coursing is discussed: 'It is a strange paradox of the media age,' he says in exasperation, 'that those with the least knowledge feel compelled to express their views most strongly – I want to get on to the councillors of Hackney and Islington in London to give them my views of inner-city link roads and creches for unmarried mothers, as I know absolutely nothing about them.'

To him there are many absurdities: 'Where the treatment of animals is concerned logic tends to fly out of the window. We all remember Sefton, the guardsman's horse injured by an IRA bomb in Kensington. But does anybody remember the names of those soldiers who were killed?

'A few years ago a horse of mine fell on the road. Both horse and rider – a pretty girl – severely injured their front teeth. The vet came within forty minutes to treat my horse, but the girl waited two days before the dentist had time to see her.

'On a television news programme, some years ago, while the Waterloo Cup was being run, the first item was a bomb outrage in Lebanon, with pictures of cut and wounded children running down the ruined streets. Second was an Irish policeman blown up by a car bomb, with his body beneath a tarpaulin and blood running into the gutter. The third item was financial; the newscaster then paused to warn viewers that they might find the next piece distressing, and two greyhounds were shown killing a hare as nature has decreed for centuries.

'Probably, however, he was correct to issue such a warning. Modern urban man is so accustomed to scenes of human violence on his television set that wounded children and blown-up policemen are acceptable facts of life. He is, however, so divorced from nature, so protected by packaging and cellophane from the realities of animal husbandry, that a hare being killed naturally is deeply shocking – as opposed to one being trapped, shot (and maybe wounded) or poisoned unnaturally, which is to him less distressing.'

Sir Mark has no time for the hypocrisy of the media either: 'Take the *Daily Mirror* for instance, which regularly castigates people for hunting or coursing. Yet when Lester Piggott won the Derby on The Minstrel it had the headline: "Superb ride from Maestro lifts home the favourite" – the jockey had hit The Minstrel thirty-one times from Tattenham Corner to the winning post. Now if Lester Piggott had got out of bed and hit the milkman's horse twice, the *Daily Mirror* would have had the headline: "Sadistic jockey beats milk-float pony" – but because it was the Derby on television and popular with millions of people, Lester Piggott was praised sky-high. It's completely illogical.'

Television's attitude is also illogical. Whenever a wildlife film shows a cheetah running down a gazelle, it is often repeated in slow motion to symphony music, with the narrator eulogising about grace, speed and beauty. Yet if a greyhound is filmed chasing a hare it is inevitably shown at normal speed, without music, as some shock-horror report about cruelty.

This in itself is peculiar, for in many wildlife films broadcast on television, predators and prey are often in totally artificial surroundings or situations, to ensure that a kill is made in front of the camera. It is then inserted into the film as an entirely 'natural sequence'. On a recent series there were examples of animals pitted against animals for the sake of 'good television' virtually every week; in media language 'good television' means that it 'entertains' the viewers. When a hare is chased by a dog it is said to be cruel and those who participate are abused; when film-makers set a falconer's trained golden eagle on to a mountain hare, and show it as a 'wild eagle', they are praised and receive prizes for wildlife photography. It is strange.

Because of coursing's contribution to the conservation of the hare and the double-standards surrounding the opposition to it, I have to agree with the 1976 House of Lords Select Committee, which was set up by the then Labour Government to report on coursing. It concluded that the amount of cruelty in coursing 'is less than 1% of the amount caused by hare-shooting', and that the ethical question should be left 'for the individual conscience and not for legislation'.

On the links between coursing, hare hunting and conservation, then I agree too with the finding of the Burns Report: 'In the case of hare hunting and coursing, it seems clear that those interested in these activities have helped to maintain habitats which are favourable to the hare and to a number of other species . . . In the case of the hare, on those estates which favour hare coursing or hunting, rather than shooting a ban (on hunting) might lead farmers and landowners to pay less attention to encouraging hare numbers. The loss of habitat suitable for hares could have severe consequences for a number of birds and other animals.'

Chapter 4

The Deer and the Forest

Deer are among Britain's most attractive wild animals and few people can fail to be moved by their grace, fine lines and sheer beauty. Thomas Bewick wrote of the red deer: 'This is the most beautiful animal of the deer kind. The elegance of his form, the lightness of his motions, the flexibility of his swiftness, give him a decided pre-eminence over every other inhabitant of the forest.' They were prized as animals of the chase, together with roe and fallow. More recently they have been joined by sika, as well as the small muntjac and Chinese water deer; exotic species from abroad that now live happily in the wild. The muntjac, in particular, is finding England a very pleasant place in which to live and is rapidly extending its range – by 10% a year.

The place of deer in the countryside is an interesting one; for although the animals are always a delight and a pleasure to see today, in the past they played an important part in the development of our landscape. As a result some of our most important and rich areas for wildlife are remnants of the old Royal Forests, where Forest Laws protected game for the pleasure of the King.

Far back into our history hunting has always been popular, but it was under the Normans that huge areas were set aside specifically for the chase, and it was said of William the Conqueror that: 'He loved the tall stags as if he were their father.' Over 80 Royal Forests were created, including such famous ones as Epping, Sherwood and the Forest of Dean. Despite being called 'forests' they included heather, heath, bog and grassland, together with ancient broad-leaved forest.

Thomas Bewick had things in perspective, by retaining a love for the deer and a respect for the chase, but disliking the system under which they flourished: 'The hunting of the Stag has been held, in all ages, a diversion of the noblest kind; and former times bear witness of the great exploits performed on these occasions. In our island, large tracts of land were set apart for this purpose; villages and sacred edifices were wantonly thrown down, and converted into one wide waste, that the tyrant of the day might have room to pursue his favourite diversion. In the time of William Rufus and Henry the First, it was less criminal to destroy one of the human species than a beast of chase. Happily for us, these wide extended scenes of desolation and oppression have been gradually contracted; useful arts, agriculture and commerce, have extensively spread themselves over the naked land; and these superior beasts of the chase have given way to other animals more useful to the community.'

Two extensive areas of Royal Forest that remain today have some of their woodland quite literally rooted back into Norman times. As a direct result Windsor Great Park and the New Forest are among the oldest, richest and most interesting areas of broad-leaved woodland left in Europe and they owe their survival entirely to their hunting past.

Windsor Great Park now covers some 5000 acres with another 10,000 acres of farmland and woodland surrounding it. Although it lies close to London, because it is directly descended from aboriginal forest, it is of great natural history importance.

It still has deer; wild roe and red that were re-introduced to a special 1000 acre enclosure. They were brought back to the Park at the suggestion of Prince Philip in his traditional role of honorary 'Ranger'. Muntjac have also colonised the whole area, even delighting in leaving the forest to eat the roses and runner beans of adjoining gardens.

Windsor Forest's antiquity can be seen in the many old oaks with huge girths, created by years of pollarding – the ancient method of tree management in which the branches were cut off as feed for deer. The old woodland is ideal for insects including some of our most beautiful forest butterflies that thrive in the glades and dappled light of a broad leaved wood. It has the white admiral, various fritillaries, the speckled wood, the purple hairstreak, and there are reliable reports, too, of the purple emperor, probably Britain's most elusive and beautiful butterfly. It has over 2000 varieties of beetle, numerous moths, dragonflies and damselflies. Fungi flourish as does the springtime forest flora with bugle, primroses, wood anemones, wood sorrel and pools of bluebells. Woodpeckers, nuthatches, tree-creepers, various tits, jackdaws and owls nest in the holes of the many ancient trees. In addition the forest is a stronghold of the sparrow hawk and hobby, and in spring and summer the whole area becomes awash with birdsong. Badgers and foxes also do well and there are tantalisingly persistent rumours of pine martens.

Fallow deer; ancient inhabitants of the New Forest.

Red deer at dawn in Windsor Great Park.

The New Forest, too, owes its very existence to its hunting past and some of its old history lives on in its Commoners' Rights for the grazing of ponies, cattle and pigs, and in the Court of Verderers. It has red, roe, sika and muntjac deer, but its most famous inhabitant is the fallow, with its attractive and distinctive summer coat. In some other parts of the country the coat does not become so spotted and in many herds of fallow there are great variations of colour from white to almost black, with many shades in between; deer numbers are controlled by stalking. Like Windsor the whole area is rich in flowers, insects, animals and birds, including birds of prey. Sparrowhawks again do well and goshawks have been seen in the area. The Dartford warbler, one of England's rarest birds, is another of the Forest's inhabitants – preferring the heathland with plenty of gorse.

Other residents include snakes, lizards, frogs, toads and all three species of British newt – the palmate, the smooth and the great crested. To help visitors see the forest's amphibians and reptiles, the Forestry Commission has a reptiliary, including the rare smooth snake, which is now doing very well in the New Forest, and the attractive sand lizard which is virtually extinct. All the pits have nets stretched over them to prevent the occupants being stolen by magpies, kestrels or people – all that is except one, the pit full of adders.

During the autumn rut. A stag roars at a challenger.

But although hunting in the past is mainly responsible for today's Windsor Great Park and New Forest, stag hunting in the present is almost totally responsible for the existence of red deer on Exmoor – another old Royal Forest – but largely a treeless one. Some of the pockets of red deer in England have grown from escaped 'carted' deer – a method of hunting which became illegal many years ago – but the red deer of the West Country are truly indigenous deer – genuine survivors. They have been hunted for generations on Exmoor, with just a brief break between 1825 and 1855 when deer numbers plummetted to about fifty Without the protection of the hunt they were killed and poached and it was not until hunting resumed that numbers again began to increase.

Today hunting flourishes, as do the deer. It is estimated that there are 12,500 wild red deer in England and Wales, 10,000 of these are found in the south west of England with 4,000–6,000 being in the countries of Britain's three packs of staghounds. Like many other country sports, stag hunting on Exmoor has suffered from deliberate misrepresentation by its opponents, and photographs of hunting 'incidents' abroad have even been sent to British newspapers as 'the Devon and Somerset Staghounds'. The usual story is that the stag is 'torn to pieces' by the hounds. This is totally false, for the stag or hind is hunted until it 'stands at bay', as shown in numerous old sporting paintings. It is then shot. As one of the joint-masters told me: 'We have nothing to hide. The stag is killed openly, in the public eye – we can't do it cruelly. The stag can't be torn to pieces either, because it's cut up into about twenty-four joints for distribution among the land owners – nobody would want a piece of venison that had been chewed-up by hounds. The incidents of hounds going in are very rare – when a yearling or a calf is trapped in a wood, or up against the wire – but that seldom happens.'

The other claim made to damage hunting is that the end is 'bloodthirsty'. This is because the carcass is bled as soon as the stag has been shot. The dead animal's throat is cut to allow the blood to drain away, in exactly the same way as an animal killed in a slaughterhouse. Any genuine indignation comes from the fact that in our modern, fast-living, plastic society, we hide death, so that when it is occasionally seen it comes as a tremendous shock, leading to feelings of outrage. Yet the same feelings experienced at the death of a deer would also be felt by anybody, who was not used to the realities of death, in a visit to an abattoir.

If stag hunting was banned on the moor (and around Tiverton and the Quantocks) it is highly likely that the deer would be exterminated in two or three years. At the moment hunting is so popular that poaching is difficult, simply because many locals have an interest in protecting the deer for their sport. Without hunting, the protection given by hunters would go, deer would be shot in large numbers for the value of their venison and one of the oldest wildlife features of Exmoor would disappear. The Burns Report says: 'Because of the widespread support which it (deer hunting on Exmoor) and consequent tolerance by farmers of deer, hunting at present makes a significant contribution to management of the deer population in this area. In the event of a ban, some overall reduction in total deer numbers might occur unless an effective deer management strategy was implemented, which was capable of promoting the present collective interest in the management of deer and harnessing such interest into sound conservation management' – a rather cautious assessment.

Deer hunting on Exmoor is very much a social and traditional event. The moor is isolated and it can be a hard place on which to live and work. Consequently hunting is important for meeting neighbours and friends, and it also gives a welcome break from

work on the farms. Today it maintains both its popularity and its importance to the life of the area. When a meet is held the rolling rural dialects of Devon and Somerset are just as prominent as the 'plum in mouth' accents of those more often characterised as the 'hunting set'. Many follow in cars to watch the progress of the hunt from the roads, and others go over the moor's tracks and bridleways in four-wheel-drive vehicles and on motorbikes. Even an ice-cream van follows throughout the winter and does good business.

As an outsider looking in, I now find stag hunting the easiest form of hunting to justify as it conserves the deer and it involves the whole community. Most outsiders seem unable to grasp the reality of hunting deer in the west country, that staghunting to the people of Exmoor is as important as football to the city-dwellers of Manchester and Liverpool.

The most unacceptable outsider on Exmoor is the National Trust which has allowed urban values and political correctness to destroy a once harmonious relationship between itself and the local community. In the early 'forties, Sir Richard Acland donated the 12,000 acre Holnicate Estate including Dunkery Beacon, to the National Trust – it was an important block of land for stag hunting. Yet in 1997 the National Trust, yielding to spurious animal welfare pressure, banned deer hunting on its land – in what was seen as a clear breech of Sir Richard's intentions. Not only was Sir Richard Acland a great benefactor – he was also a great supporter of the whole of Exmoor's traditions and culture.

It is significant that in other areas the National Trust has actually bought land with a built –in guarantee to allow hunting to continue – the high profile Snowdon purchase with its apparently hushed-up hunting agreement is one case in point.

The deer hunting ban followed research by Professor Patrick Bateson of King's College, Cambridge. I, as a layman was astonished by Bateson's report – in it he referred to 'Bambi' – in places he seemed to rely on supposition rather then science or observation, and to me political correctness appeared to take precedence over fact. Yet the National Trust accepted the Report – only to see it largely discredited by fellow scientists two years later following more research by Professor Roger Harris and colleagues, known as the Joint Universities Study on Deer hunting. The Burns Report too seems to arrive at some vastly different conclusions to those reached by Bateson – yet in the Queen's Birthday Honour's List in the summer of 2003 Professor Bateson and the National Trust's Chairman at the time of the deer ban, both received knighthoods.

For an outsider to say that hunting is cruel and should be stopped, when it is hunting that has preserved the deer in the first place, is both arrogant and an intrusion. Some local people do not like hunting, of course, but the most vociferous objectors are those from south-eastern suburbia who have chosen the south-west in which to retire. Perhaps they should think of other things; how their interference in the property market has pushed up prices so that young local people cannot afford to buy houses in the communities in which they have been brought up — something more immoral in my view than chasing a deer. In any case, those people who have taken their suburban values and sensitivities with them to Exmoor, would be better advised to move to more suitable areas for them, such as Hove or Worthing.

I visited a farm in the middle of Exmoor during the autumn rut of the red deer. Despite rain it was a good time. Next to the farm is a deep combe with the Danesbrook, a tributary of the River Barle, running at the bottom. A buzzard circled, rooks called and from oak woods across the valley came the roaring of a stag. Then, closer to hand among gorse, came a reply. There were several stags in the gorse, their antlers merging well with their surroundings. One in particular had a fine head; he winded us and moved off towards a small group of hinds. The red deer on Exmoor are far larger than those of Scotland as conditions are less hard, with plenty of food and shelter.

We walked back by the 'brook', with the music of water over rocks accompanying the cracking of twigs and the squelching of boots on sodden ground. Oaks, alders, birch and beech gave shelter, and a carpet of moss padded sound as more deer ran from us. The valleys, woods and streams of Exmoor are ideal for nesting dippers, wagtails, redstarts and pied flycatchers, and in the autumn salmon 'run' upstream to spawn. On the higher hills a few red grouse still survive and wheatears, meadow pipits, stonechats, whinchats and nightjars are other birds to be attracted by the heather, bracken, gorse and grass. Buzzards, kestrels and sparrowhawks find good hunting over the mixture of farm and moor, valley and heath, while in the winter they are joined by merlins and hen harriers.

John Pugsley was keen for me to see his part of Exmoor. He farms 1300 acres, most of it as a tenant, with 2500 sheep and 300 cattle. His land consists of many habitats – improved grassland; ancient heath; bog and wooded combe. He has added to this by planting several small areas of woodland and creating pools in a narrow damp valley. Foxes cause problems during lambing and some lambs are always lost. He suffers from deer too, but he loves to see them, as long as the damage is not too great. He hunts foxes, not deer, but welcomes the Devon and Somerset Staghounds on his land: 'One of the most important factors of hunting is that it breaks up the large herds into little groups, and half a dozen deer in a field are more acceptable than 60 or more. Those who criticise hunting seem to be unaware of this. Certainly if hunting stopped, the deer pressure on my farm would become unacceptable and I would have to shoot. With the price of venison being what it is and the size of the Exmoor deer I believe they would be shot out within three years. It would be a great pity because both the deer and the hunt are part of the moor. Even now I can't grow roots for my sheep because of the deer. They're very choosey eaters and like nice, sweet things; they only choose the best.'

Because of his genuine affection for deer and their place on the moor, he resents interference from outside, whether from 'newcomers', pressure groups, or politicians. 'They don't understand,' he says. 'To make matters worse the media are on the side of the "antis" and distort the truth. I expect, too, that we will be increasingly pressurised by

Rodger McPhail – red deer in the Highlands.

the hooligan element, again from outside. If ever they have their way it will be a sorry day for the deer and for Exmoor. As it is, the deer hunting on Exmoor has helped preserve the appearance and the wildlife of the moor; without them even larger areas would have been ploughed or covered with forestry. The preservation of the red deer has helped conserve the whole character of Exmoor for the benefit of everybody.'

The importance of hunting in the conservation of Exmoor was confirmed by Lord Porchester in his 'A Study of Exmoor', in 1977. As inspector at the enquiry he was asked to write the Report for the then Labour Government because of the various pressures on the moor that were threatening to change it beyond recall. Of the deer he wrote (Chapter 5 paragraph 4):

'Special mention must be made of Exmoor's red deer. The fortunate visitor may catch a glimpse, and a fine sight they make. They live on Exmoor in good numbers and are believed to be genuinely wild, descendants of the indigenous population of the Royal Forest. Stag hunting and fox hunting are both notable features of the Exmoor way of life. Whilst it may come as a surprise to some and prompt the wrath of others, it is undeniable that stag hunting on Exmoor operates as a force for conservation. The stag hunt is supported by almost every member of the farming community and this guarantees the deer's continuing existence. In the normal run of things, they can do considerable damage to crops and, without the active participation of the farmers in the hunts, their days would be numbered. Moreover, such is the local interest in stag hunting that substantial areas of land within the Critical Amenity Area are corporately owned with a view to securing the deer's habitat. So long as stag hunting continues it is unlikely that such land will be substantially altered by conversion or enclosure'.

Away from the old Royal Forests, deer are now doing surprisingly well throughout Britain. This is a genuine surprise, for in the eighteenth century Bewick was writing of red deer: 'In the present cultivated state of this country, therefore, the stag is almost unknown in its wild state. The few that remain are kept in parks among the fallow deer.' The roe deer was also apparently scarce: 'The roe was formerly common in many parts of England and Wales; but at present it is to be found only in the Highlands of Scotland.' Fallow were slightly more numerous, but most were confined to parks, such as Petworth, in Sussex, which is still a deer park today, for their beauty, and for their venison.

Things have changed. Now, in addition to Scotland's large red deer population, there are herds in several parts of England, including the south, south-west, East Anglia and the Lake District. Roe, too, are numerous in Scotland and are increasing both their numbers and their range in England. Fallow are also doing well in England, with a few herds being found in Scotland and Wales as well. The reason for the growth of the overall deer population is almost certainly linked to the increase in forestry.

The roe has probably benefited most from forestry and is the smallest and most attractive of our native deer. Its range covers much of Europe and the Mediterranean, where its beauty has been praised for hundreds of years. 'The Song of Solomon' in the Old Testament, is one of the most beautiful love poems ever written and contains several references to the roe.

Unfortunately, in the early days of the roe's revival, their beauty was not seen, just the damage they caused to trees and crops. 'Deer drives' were held by farmers and foresters, using shot guns of insufficient calibre, and a tremendous amount of suffering and pain was caused as a direct result. Even those who shot pheasants on forestry land were encouraged to shoot any deer seen as 'vermin'.

Fortunately it was the deer stalkers who came to the rescue of the harassed and shot-up deer. They pointed out that deer were a valuable resource, providing venison as well as a stalking income. Furthermore, by studying the whole ecology of the roe and fallow they showed that some deer could be tolerated and deer damage kept to a minimum. The move towards toleration was helped when the use of shotguns against deer was outlawed. Now most deer are killed by using high velocity rifles, and instead of spending money on 'vermin' control, the Forestry Commission receives good money for letting its sporting rights, a sizeable proportion of which comes from stalking, with more income accruing from the sale of venison itself.

Now with the current problems facing the farming industry, roe and fallow stalking may well form a welcome source of additional income on many farms. For just as the deer have moved onto forestry land, they have also been attracted to the covers and shelter belts left for pheasants and foxes. This again shows the benefit to many forms of wildlife from a healthy and varied countryside.

The stalking of deer is highly skilled and the use of high powered rifles with telescopic sights means that death is nearly always immediate and humane. A deer population left untouched would cause considerable problems; damage to farm crops; the prevention of natural tree re-generation, and eventually, with no wild predators, over population. Deer also become involved in, and cause, traffic accidents and according to the Burns Report there are 40,000 deer related traffic accidents in Britain a year. Proper management keeps the population at an acceptable level and damage to minimum, and the British Deer Society has done much work on the ecology and management of our native deer.

To shoot deer properly is not an act of cruelty, but a skill, in which an understanding and a respect for the quarry is demanded. People involved in country sports are a constant surprise to me; one of them, John Hotchkis, the composer, was best known for the music he composed for the Consecration of Coventry Cathedral in 1962.

In addition he wrote numerous film and television scores; he wrote the 'Fanfare for Europe' at Coventry Cathedral in 1973, he composed works for Baltimore Cathedral, an

The red squirrel in old Caledonian pine – its natural habitat. Well planned forests can make allowances for deer and other woodland life.

A broad ride in Sitka spruce – not designed for wildlife.

overture, chamber music and much more. Yet he was also involved with deer management and conservation for many years. He was a quiet, unassuming man, yet he saw nothing odd about his love of music, pastoral living and shooting deer. 'I have an enormous affection for deer and fallow are my favourite,' he said. 'They are a beautiful part of the countryside. My love for deer, music and the country are in no way contradictory. To stalk deer properly is humane and it is our duty to fight cruelty whether to animals or humans. Deer management is done for their own benefit; without it numbers would become too high, farmers and poachers would take things into their own hands and it would be unselective and cruel.

'Roe deer I find particularly attractive, they hold on to small pieces of woodland which are not too dense and where there is little disturbance. If there are suitable corners of woodland and adequate food, then the farmer or forester will get little damage. It is the same in Scotland. Red deer, forestry and farming should be able to co-exist quite well with proper management. But that means the forests have to be properly managed as well. Too often in the past it has been bad forest design that has actually caused deer problems. The fenced, serried ranks of pines mean that the deer have nowhere suitable to go in winter, and so they go on to farmland. A well designed forest allows deer in for

OPPOSITE: *The magnificent Capercaillie (by A. Thorburn) – its numbers are now steadily falling.*

Capercaillie (♂ ♀)

shelter and feed when the trees are large enough, with deer lawns, glades and large rides which give them grazing away from farm fields and new plantations.'

His mention of Scotland's deer and the potential conflict with forestry was inevitable. Most people regard the Scottish highlands as the natural home of the red deer. But it is not; the deer simply retreated there, away from the increasing pressure of farms and the felling of forests. Now they find refuge in the highlands, where food is poor and conditions can be hard, and so the Scottish red deer are far smaller than their relatives in the south. An average Scottish stag will weigh about 200 lb while a stag from England will average about 400 lb. Unfortunately the new Scottish forests are alien too, with every effort made to achieve maximum timber production with as little deer damage as possible.

This is a lost opportunity, for well planned forests can make allowances for deer and other woodland wildlife. There were about 350,000 red deer in Scotland in 1987; The Red Deer commission has become 'Deer Commission Scotland' and with the change of name there has come a degree of vagueness; 'there are possibly between 200,000 to 400,000 roe deer, but all the figures are vague and almost certainly inaccurate.' Percentage culling is also a thing of the past with deer control concentration on special areas to prevent damage to woodland, the national heritage and to ensure public safety.

The traditional landowner in Scotland views deer as a sporting asset for stalking; the forester, whether private or public, still tends to regard deer as 'vermin'.

The Deer Commission Scotland is the statutory body set up to further the conservation and control of red deer in Scotland and works closely with other government departments under the Scottish Parliament and other conservation bodies.

At one time much of Scotland's wild venison was exported to Germany. Since 'foot and mouth' that market has contracted as the Germans now look to Russia and Eastern Europe for their venison market – but fortunately the home market has greatly improved.

The Deer Commission Scotland has also become increasingly keen to encourage those people planting new forests to change 'forest design', to lessen the conflict between forestry and deer. This means 'external design', to allow deer to move easily down into the valleys for shelter during the winter, without having to break through miles of preventative deer fencing. 'Internal design' is also important to create open spaces to permit deer into maturing forests without doing damage. Not only do they provide deer with harmless feeding areas, but they also make the forests more attractive aesthetically, and for other forms of wildlife. Birch trees for winter browsing and planted turnips can all help to draw deer away from areas where they can cause problems.

Rothiemurchus is a fine example of a private estate which is run to combine the needs of traditional sport and wildlife with commercial forestry. In addition, it views tourism as another legitimate form of income, with walks, tours, shops and a trout farm. It has also moved into deer farming. Because of its mixed interests it has maintained a variety of habitats and possesses areas of traditional highland Scotland, complete with red deer, heather, golden eagles and ancient Caledonian pine forest.

The 22,000 acres of the estate lie between the River Spey at Aviemore and the Cairngorms. Rothiemurchus has been in the Grant family since 1588 when it was owned by Patrick Grant of Muckerach, second son of Chief of Grant. Today it is owned by John Grant, Younger of Rothiemurchus. Much of it has been designated a Site of Special Scientific Interest by the NCC and forms part of the newly created Cairngorm National Park. Lower down, around the farm buildings, are small grass meadows for cattle, together with arable fields to provide barley for cattle feed and the distilling of whisky. Birch woods abound too, which in many parts of Scotland are in decline due to overgrazing.

Climbing away from the floor of the Spey valley there is pure Caledonian forest. The old Scots pines are magnificent trees, growing to as high as 120 feet. Beneath them are juniper bushes, bilberry, mosses and lichens. It is a tragedy that the Forestry Commission has not made more economic use of our natural forests, for the mature Scots pine produces excellent timber.

The forest is of great importance as a sheltering area for red deer in winter; but numbers are kept at a proper level and there are many young trees emerging, showing that natural regeneration is taking place. This will ensure the continuance of a healthy forest into the future. The habitat created by the old pines are ideal for red squirrels, roe deer, crested tits and crossbills; all finding their own special niche.

Blackcock numbers are still quite good, too, whereas they are in sharp decline in many other parts of Scotland and northern England. Capercaillie are to be found, but numbers are falling, probably because of disturbance by walkers and tourists, some with dogs running loose, which can lead to the loss of nests during the breeding season. It is as if some visitors see the area in the same way as they view a town park; it is simply a place for enjoyment. They have no knowledge, or concern, for the wildlife whose territories they are invading. Another problem is that of 'naturalists' seeking out the 'lek' in the spring, when the magnificent cock birds display to impress the females and each other. Again, disturbance, even by people just wanting to see the birds, at a crucial time in their annual cycle, can lead to a fall in breeding success. It is a case of free public access, which is allowed by law in Scotland, causing a problem to wildlife. Deer fences and predation by the protected pine marten are yet more problems faced by the capercaillie.

Higher, above the tree line, there are the heather moors and high tops of the Cairngorms – one of the last great wilderness areas of Britain. In summer the red deer go far off into the hills, to avoid the heat and flies of the valleys. Among the heather there are grouse and mountain hares, and curlews, golden plovers and dunlin all breed. Golden eagles, buzzards and peregrines hunt, creating dangers for the young and unwary. The summits form the homeland of the ptarmigan, where the rare dotterel also lays its eggs. Incredibly, 'twitchers' – over-keen birdwatchers – also arrive on the

Blackcock at 'Lek' – the traditional displaying site.

tundra-like tops looking for dotterel and some have even walked along in line, in similar fashion to pheasant beaters, to flush the birds and record a tick for their bird species lists. It is almost as if listing birds has become a substitute form of 'train spotting' for some peculiar and obsessive bird watchers.

There are other species breeding on the estate; ospreys, goldeneye and probably even greenshank. It is one of the richest and most extensive relics of old Scotland, giving an exciting view of our wild, highland heritage, of which John Grant is rightly proud. It is there because of 'sport'. Without an interest in stags, grouse and ptarmigan the estate would have become 'commercial' years ago, leaving today's visitor to see only dense pine plantations and high barren tops, over-grazed by sheep; hardly reasons for visiting the Highlands or the Spey valley.

Chapter 5

The Pheasant and the Copse

To many country people the sight of pheasants feeding on stubble, or flying over a snow covered winter landscape, captures the complete spirit of the traditional English countryside. It is an inaccurate picture, for although the pheasant has been part of the country scene for several hundred years, it is in fact an imported alien, that has gradually colonised most of Britain, from a mixture of escaped and released birds. Now each year the truly wild population is augmented by the release of twenty-five million reared birds, bred specifically for shooting; for pheasant shooting is the most common and popular form of game shooting among Britain's 700,000 shooters – representing 90% of all quarry shot. Surprisingly this twenty-five million birds represents a substantial increase in the number of birds reared since the first edition of *The Fox and the Orchid*; the number then was between ten and fifteen million. There are two reasons for this increase: firstly the drastic decline in farm incomes which has meant that more farmers now let their shooting in an attempt to increase their incomes. Secondly: at the same time that farm incomes have dropped – city incomes have risen, leading to an increase in social shooting. Shooting is again being seen by those successful in business or the 'city' as a status symbol. The word 'city' has a double meaning as many of these status symbol shooters are also city dwellers.

The various species of pheasant originally inhabited an area of Asia from the Caucasus, in Russian Georgia, eastwards into China, living in marshes and woodland.

The 'Ring-neck' pheasant.

The first introductions to England came from the western part of their range and became known as Old English pheasants. Later birds originated from further east, and because of the distinctive white ring on the neck of the cock, were called Ring-necks. Other races, too, have been introduced and they all interbreed freely. Nevertheless some of the old features can still be seen and the characteristic markings of the Old English and Ring-neck continue to be common.

There is some disagreement about the date of the first arrivals. Pheasants were certainly popular with the Romans as table birds, and it is highly likely that they were reared for eating during the Roman occupation of this country. The first British record dates back to 1059 and it seems probable that several introductions were made soon afterwards by the Normans. With the development and growth of shooting as a sport in the eighteenth and nineteenth centuries many more were brought in and released during that later period.

In my childhood truly wild pheasants were common on the farm and every year we would eat several during the course of the shooting season — each one preceded, of course, by 'light pudding'. Most were shot by old Jim, with his single barrel twelve bore and whenever he was ploughing or working in the fields he would take his gun, and Peter, the farm spaniel, with him – 'to take them for a walk' during his dinner hour. He would wander along the ditches, by the brook, or through the spinney, hoping that Peter would flush a bird so that he could have a shot. If ever he saw a cock pheasant when his gun was not to hand, then his language would almost match the ferocity of a cartridge. I even saw him throw his cap on the ground in frustration. On a good day he called them 'old long tails' – and just as he enjoyed shooting them in the winter, nothing gave him greater pleasure in the summer than seeing a female pheasant with her young brood, flying over the corn. (The extraordinary collective noun for a family party of pheasants is a 'nide'.) If he could, he would cut round a nest in a field of hay, but if he did accidentally destroy one, he would save any undamaged eggs to place under a broody hen. From one such brood only a solitary chick survived; it grew up to be a family pet – a magnificent cock we called 'Charlie'. One day it escaped and my younger sister was very upset. Jim tried to comfort her by saying that Charlie would be back in a couple of days. He was right, for two days later while he was taking his gun and Peter for their dinner-time walk, he saw a cock pheasant perching in an elm. Jim was not a man for 'sporting shots' and he blasted it where it sat. It was Jim's turn to be upset, for it was Charlie, and we ate him for Sunday dinner. Poor old Jim was not a typical pheasant shooter, however; shooting with beaters, and driving pheasants towards a line of guns, is what most people regard as 'pheasant shooting'. It is done both for enjoyment, as good shooting demands a high degree of skill, and often as a form of income for the landowners concerned. Because organised shooting involves the death of many beautiful birds, it inevitably arouses opposition. As a result, the feelings of some people are best summed up by the words of Alexander Pope (1688-1744) in his poem 'Windsor Forest':

> See! from the brake the whirring pheasant springs,
> And mounts exulting on triumphant wings:
> Short is his joy, he feels the fiery wound,
> Flutters in blood, and panting beats the ground.

Unfortunately, emotion quite often gets the better of reason, as there are many contradictions in the arguments ranged against shooting. The obvious objection is that pheasant shooting is cruel; but undoubtedly the 'cruellest' form of shooting is pigeon shooting. There, artificial decoys are used to lure the birds within range of the guns, and dead pigeons add to this deception. Wounded birds can be left flapping where they fall for hours until shooting stops. Despite this, I have never heard an opponent of 'blood sports' speaking out about the 'evils' of pigeon shooting, yet each year I hear pheasant and grouse shooting attacked and ridiculed. It can only assumed there is a strong political element within the various anti-blood sports movements and that attacks on selective forms of shooting are seen as an extension of the 'class struggle'. Pigeon shooting is on the bottom rung of the shooting ladder, with many 'working class' participants, and so it is ignored.

There is also considerable opposition to the idea of rearing and releasing pheasants to be shot, although the birds are mainly released well before the start of the shooting season. Some shooting men themselves have misgivings and prefer to shoot completely wild pheasants. Again much of the opposition and doubt is misplaced. The pheasants shot, whether reared or wild, are eaten; either by the participants themselves, or by the general public, as many thousands of pheasants are sold to game dealers each year, so ending up in ordinary butchers' shops. It is ironic that whereas the most humane method of rearing hens is considered to be 'free-range', the rearing of 'free-range' pheasants is considered to be unacceptable. During the course of their lives the pheasants are protected, fed, and pampered. They are content, and become both free and wild. Then one day in the autumn or winter, a small percentage of them will die in the woods that they know and regard as home. Most of those that are wounded will soon be retrieved by dogs – usually labradors or spaniels that have been bred specifically for the purpose. About 40% of released birds are shot – the rest have to face the natural hazards of disease, predators, hunger and cold. If pheasants are unshot and wild, on unkeepered land, then the annual mortality rate is 50 to 70%, or even more, as life away from the guns is more dangerous. The low survival rate explains the large clutches of eggs layed by the hens each summer; sometimes with 'dumping' by other hens the clutches can contain fifteen eggs or even more – the record is twenty-eight.

To me, the reared pheasant is far more acceptable, and moral, than battery hens or broiler fowl – kept in huge numbers simply to provide cheap eggs and meat. The supermarket shoppers never complain, as they simply see them as attractively presented 'products' – like the canned peas or bags of oranges. The shooting of a wild pheasant is seen as murder – the death of a broiler fowl is not seen at all. One sight that disgusts me each year is of lorries loaded with 'clapped-out' live battery hens, being carted down motorways in draughty crates; they are destined no doubt for meat pies or cheap restaurants; a fate the pheasant manages to avoid.

The other main argument against organised pheasant shooting is that the preservation of pheasants, and the rearing of others for release, involves the 'destruction of wildlife'. In years gone by this was probably true, as some gamekeepers were fanatical in their persecution of anything that posed even the slightest threat to their birds. As a result all birds of prey, including owls, were shot and trapped, as well as all foxes, stoats, weasels and hedgehogs. Today most gamekeepers take a more enlightened view and the main predators to be controlled are foxes, stoats, weasels, rats, mink, crows and magpies. In my view some keepers still kill too much 'vermin' and no doubt there are a few who continue to kill protected species such as tawny owls and sparrowhawks. I believe that there is a case for controlling, not exterminating sparrowhawks, but with some government bodies and conservation organisations refusing to listen to properly presented and rational arguments it seems to me that at times gamekeepers can almost be forced into illegal action. But things are changing, and with reared birds a higher level of predatory creatures can be tolerated. Even so the continuing use of the words 'vermin' and 'vermin control' indicate that there is still some way to go. With our increasing knowledge of

'Hedgerows and corners of fields which have been left rough.' Pheasants need a varied habitat.

predator-prey relationships, the word vermin should have disappeared long ago – keepers and shooting men should be talking about 'predators' and 'acceptable population levels', part of their job is 'population management' which is just as important as habitat management.

It also has to be said that some of those who criticise predator control are not always honest. I have often queried with gamekeepers the need to kill hedgehogs, as I like to see them. There is a difference of opinion; some 'keepers claim that they create a constant problem; others say that they can be tolerated. It could well be another case of the individual habits and preferences of the hedgehogs themselves. Because of this, some years ago I asked the RSBP for its view on gamekeepers killing hedgehogs. The RSPB spokesman was scornful and dismissive: 'It is ridiculous,' he said, 'a hedgehog's jaws are not strong enough to bite through an egg shell and they do no damage to ground nesting birds. Some gamekeepers just like killing things.' Sixteen years does not show the RSPB in a very good light. Hedgehogs were introduced to a number of Hebridean Islands, by gardeners wanting to control slugs naturally. Sadly the hedgehogs preferred other morsels – the eggs of not only small birds, such as snipe, redshank, ringed plovers, terns, dunlin, skylarks and corncrakes, but also much larger birds including wigeon, shoveler, teal, gadwall, fulmer and black guillemot. The hedgehogs thrived – the population it is believed has grown on South Uist alone to over 3000 and serious conservation damage is being done. And which organisation is joining the chorus to move or cull the hedgehog? The RSPB of course. Suddenly those 'weak hedgehog jaws' have become strong. Sadly, the RSPB seems to be too fond of fence-sitting to publicly support a cull – presumably for fear of offending some of their more whimsical members. So although privately RSPB officials want a cull as quickly as possible – publicly they will only say: 'We want the issue resolved successfully and of course it is the responsibility of Scottish National Heritage.' It would appear from this that the RSPB may be putting fence-sitting and subscription renewals before real conservation.

From the evidence available it would seem that the RSPB's views on the damage done by some birds of prey such as the sparrowhawk, goshawk and hen harrier is as inaccurate as its old views on hedgehogs – yet for the sake of conservation and political correctness it reels off the same old inaccurate fantasies – 'birds of prey do no damage to conservation and endangeared species'.

Morally there seems little difference to me between a wildlife warden killing foxes and rats to save attractive or rare species of birds, and the gamekeeper preserving pheasants. The most important point however is that pheasants need a healthy and varied habitat before they can flourish, which is also ideal for many other forms of wildlife. Most of the birds, butterflies and insects which benefit from the habitat created do not conflict with the pheasant shooter at all, and can actually be of benefit to young game birds and to the farm crops as well. Because of the rich surroundings created by shoots, it is hardly surprising that many of Britain's leading naturalists, past and present, had their interest in wildlife fired by a liking for shooting. The list includes Gilbert White, Charles Darwin, W. H. Hudson, Richard Jefferies and, more recently, Sir Peter Scott, founder of the Wildfowl and Wetlands Trust. It is also true that as interest and knowledge grew with some of them so the desire to watch and learn replaced the desire to shoot. This was the case with Sir Peter Scott.

Another great but current conservationist/shooter is Norfolk farmer Chris Knights – Vice Chairman of The Countryside Restoration Trust. Yes, he does enjoy pheasant shooting – which is part of the culture of North Norfolk, along with coursing – but his farm is also a wildlife haven. He has barn owls, lapwings, grey partridges, skylarks yes – and sparrowhawks, and he has the highest density of breeding stone curlews in the country. In addition he is one of Britain's top wildlife photographers, winning The Bird Photographer of the Year award in 2003, given by *British Birds*, for an outstanding picture of a skylark – taken on his farm.

Today the pheasant is in no danger of becoming extinct. Although in the over-intensive prairie lands it is extremely scarce, it thrives in areas that have been planted to give it cover, or hedgerows and corners of fields that have been left rough, especially for the benefit of 'long tails'. It is there too that other birds, flowers and butterflies still flourish, giving continuing glimpses of the 'traditional British countryside'.

I enjoy watching pheasants; sometimes in early May as I wait near a fox earth, hoping to see cubs, I will be entertained by a cock strutting, crowing and flapping his wings in a mixture of vanity and challenge. In his usual style, using no punctuation, John Clare made an interesting observation about this: 'It has been supposed by naturalists who are more fond of starting new theories than proving old facts that our yard fowl the cock was originally from the wild pheasant, but I think this is merely a wide supposition as the yard cock always claps his wings and then crows while the pheasant cock always crows first and claps his wings afterwards.'

Then as night falls, birds will clatter up to roost in the high overgrown hedge of the brook meadow. They like the hedge; there are patches of brambles and grasses that provide good nest sites; it gives ideal shelter, and its rich autumn harvest provides abundant food. Hawthorn, ash, dogwood, blackthorn, buckthorn, elder, wild rose, ivy, bryony, blackberry and wild hops yield a wide variety of berries and seeds. I have seen greenfinches, goldfinches, bullfinches, linnets, yellowhammers, fieldfares, redwings, thrushes, rooks, magpies, blackbirds, kestrels, cuckoos, robins, wrens, blue tits, great tits, long-tailed tits, spotted flycatchers and many more using the hedge, and every year it contains many nests. One summer, much to my amazement, I even saw a parrot; it stayed in the area for three or four months, before catching cold, or flying off in the forlorn hope of finding a mate. Butterflies also do well; the large white, small white, green veined white, common blue, red admiral, painted lady, comma, small tortoiseshell, peacock, gatekeeper, brimstone, small skipper, Essex skipper, large skipper, ringlet, small heath and meadow brown – I have seen them all in and around the hedge. None

Perfect Dorset pheasant cover – especially planted by Christopher Pope.

of the species is rare, but they simply cannot be found on farms dominated by the sprayer and flail mower. The presence of butterflies is living proof of a healthy environment and that the farmer has cared for his land. This simple truth was confirmed for me again in 2002 when I undertook the Great British Butterfly Safari. In it I attempted to see every British species in a single summer. I managed it – just – notching up fifty-nine species of butterfly.

The importance of the hedge was shown to me dramatically in the November I was originally writing this book. I was walking over nearby hedgeless land; the fields were cultivated so close to the bank of the brook that I had to walk on the crop itself. The brook banks had been flail-mowed and bird life was virtually non-existent, apart from a few lapwings standing on the ploughed land. It was depressing. High above, a flock of field-fares came into view, the first I had seen of the winter. Suddenly they saw our brook meadow with its sprawling hedge – they veered away and fell like falling leaves to feed hungrily on the rich crop of haws. As I reached the meadow a little owl flew from a hawthorn tree and Bramble flushed two pheasants from our overgrown section of bank and then he put out a fox. In two hundred yards I had seen more than in the previous mile.

The pheasants often make use of the grasses, reeds and rushes of the brook for food and shelter. Once I remember watching two cock pheasants from a hide at the RSPB's reserve at Minsmere, in Suffolk. They were fighting and displaying. I was surprised to see them in the reeds at the time – although the surroundings were possibly very similar to their Asian homeland.

The dramatic link between pheasants and general wildlife can best be seen on land where there is a definite management policy to provide birds for shooting. This was shown most clearly to me on a 1000 acre farm in Suffolk, part of the Hiam estates. There John Wilson, a grandson of Sir Frederick Hiam, has changed over from a system of rearing and releasing pheasants, to an all wild pheasant shoot. But his interest in the life cycle and welfare of his pheasants has grown into a deep and genuine knowledge of wildlife, and at times I had to remind myself that I was looking at a commercial farm and not a nature reserve.

He grows wheat, barley, sugar beet, potatoes and parsnips. In addition he has a hundred acres of grassland, mostly water meadow and just over a hundred acres of woodland. His land has twenty-five miles of hedgerow, as field boundaries. He grows plenty of nettles and thistles, too, and many of his farm roadways have a pleasing natural look.

The farm is in the heart of East Anglia's prairie farming belt. But whereas one of his neighbours has fields of over a hundred acres in size, he has kept his field size down to an average of about twenty-five acres. He believes that good habitat, with plenty of insects provides a 'biological control' for pests: 'A one hundred acre field is too big as the beneficial insects cannot get to the middle from the edge. Anything over forty to fifty acres is too big.' The size of his fields enables predatory beetles to cover the whole area, particularly ladybirds and ground beetles that eat harmful aphids. Some of his neighbours have to spray for aphids every year, but for him severe infestation is now rare. When he does spray he does not spray automatically, as a precaution (a bad and expensive habit encouraged by the agro-chemical industry), but only when and if he has a problem developing.

His army of aphid eaters has been helped by the fact that around some fields of cereals he leaves unsprayed headlands; this allows a limited amount of 'weed' growth and causes a large increase in the population of beneficial insect life.

He will persist with a number of unsprayed headlands as long as farm economics permit. One sprayed headland, growing parsnips had a very interesting weed – henbane; it was good to see John Wilson point it out to his workman, rotavating, in order to avoid it. It is an unusual and interesting plant and was used by the notorious Dr Crippen to poison his wife.

The hedgerows around his fields vary. At one time, hedges cut into an 'A' shape were considered to be best for game and wildlife; he disagrees and now allows many of his hedges to grow between 6 and 8 feet high and 2 and 3 feet wide, and where possible he likes to have a 3 feet wide grassy bank on each side. These allow many insects and plants to benefit from the hedgerow; they also make good, dry and safe nesting sites for game. If the grass has to be cut, only the seed heads are removed – the banks are not mown smooth, so leaving protective cover for nesting gamebirds.

To prevent spray drift damaging the hedgerows, or harmful weeds – rough stalked meadow grass and sterile brome – growing into the fields, some arable crops have a 'sterile strip' of bare, rotavated soil dividing the crop from the hedge. This has another advantage of giving pheasant chicks an area in which they dry off after rain, so greatly improving the survival rate of young birds.

Other hedges are allowed to grow, to be 'coppiced' (cut down to ground level) at intervals of between 15 and 20 years. This stimulates growth and provides good firewood. Newcomers to country life often hysterically mistake this form of good hedge management as hedge destruction. Their embarrassment and education start when the hedge begins to grow new shoots the following summer. The roadside hedges are cut into a 'half A' shape and a few hedges have been let go completely to provide winter berries and shelter for wild birds. His hedge cutting is never done in the birds' nesting season, or immediately after harvest, but after the seeds and berries have been eaten. During the cutting operation suitable young saplings are left to become hedgerow trees, and any gaps are planted with young oaks.

In addition there are over a dozen game spinneys and shelter belts of both softwood and hardwood, to provide cover and nesting sites, as well as to make the pheasants fly high during the shooting season. One ten acre area of woodland is to be coppiced in rotation – one of the oldest forms of tree management and very beneficial to woodland flora and butterflies.

Water and the vegetation in it and around it are very important to wildlife. John Wilson has twenty-two ponds on his farm – providing suitable conditions for fish, frogs, toads and two types of newt (the common and palmate). Sixteen of these ponds have been incorporated into game covers and areas kept for wildlife, and six have been left within arable fields, despite the inconvenience. In his unploughed water-meadows he has tried to control water levels to create real wetland, and by the Black Bourne river he is trying to restore an old reed bed. Some of the river bank looks like pure jungle with tussock sedge, Himalayan balsam, reedmace, yellow iris, comfrey, marsh marigold, hemp agrimony and assorted rushes and reeds; it is a paradise for sedge warblers, reed warblers, moorhens, water rails and pheasants. It would break the hearts of tidy-minded flail-mower drivers. However, that possibly would not be a bad thing.

The sum total of John Wilson's sympathetic farming is good crops, an attractive landscape, a fine pheasant shoot, one of the richest wildlife areas in south-western Suffolk and a very pleasant farm on which to work and live. He has recorded 120 bird species on his land, as well as an excellent twenty-four types of butterfly, at least nine dragonflies, four damsel flies and two varieties of orchid – the bee and the early purple. Kestrels, sparrowhawks and tawny owls all breed on his land, with hardly any problems. Occasionally there will be a rogue tawny that kills pheasant poults, but that can be tolerated. Goshawks are also in the area; again this does not worry him as they are such attractive birds. Hen harriers are seen during the winter and will take the odd English partridge, but this is of more concern to the partridge than to John Wilson. Interestingly the harriers nearly always take the skulking and walking English partridge instead of the running Frenchman. Recently a barn owl was seen flying through the farmyard once more; he hopes that it will return with a mate to breed.

One of the most attractive breeding birds on the farm is the kingfisher, which nests in an old sandpit some way from the river. When we were looking at one hole, thought to be the nest, the bird flew from another. The river has a good fish and eel population and, due to this and the attitude of John Wilson and two of his neighbours, his farm became the first site for a successful otter re-introduction project in 1983. Under the supervision of English Nature and using otters reared by Philip Wayre at The Otter Trust, near Bungay, two bitch otters and one dog were successfully released into the wild. Some time before, a 'carp pond' had been established in the water meadows for fishing; it has now been re-named the 'otter-larder', but the new fishermen are very welcome. It is believed that from this successful release – together with one or two more – virtually the whole of East Anglia has been successfully recolonised including the little brook on the southern boundary of my farm.

As if all this was not enough, when gravel digging took place on John Wilson's land in 1985, to provide material for a new road, he agreed, only on condition that the ten acres of water and workings were left in a suitable condition to give maximum encouragement to wildlife. As a result gently sloping shores were created as well as shallows, and even a cliff to encourage sand martins. He then sunk sixteen 'big bales' into the water for the benefit of bacteria and bugs, which in turn helped plant growth and fish life. Now little ringed plovers breed on the gravelly shore and there are also breeding mallard, tufted duck, gadwall and little grebe. He has successfully put up heaps of bales for shelducks to nest in, and sand martins have found the cliff excavated for them. Near the water he has planted alder buckthorn (for brimstone caterpillars), wayfaring trees and the guelder rose. With maturity, in several years time, it will be a wildlife paradise. Over the last five years he has planted 22,000 trees, of eighteen species, on his farm. When creating and planting water and woodlands the work never stops; young woods have to be kept weed free; old woods have to be cut back to allow in light for birds and ground flora, and ponds and gravel pits have to be prevented from choking with weeds. This costs time and money; but it benefits wildlife and game, and it is a price that many shooting men are prepared to pay for their sport and for their surroundings.

John Wilson believes in a degree of public access and has created a number of walks and vantage points. But he regards unlimited access as unreasonable as it could result in damage to his crops, interference with breeding game and disturbance to his various wildlife schemes. School parties and other bodies visit regularly, however; I would like to see him cater for groups of prairie farmers and anti-blood sports campaigners – both would learn a lot.

To see a shoot that still relies largely on reared pheasants I travelled down to Dorset, to a farm just outside Dorchester. I had a few reservations before making the journey. Like many people I have come to associate rural Dorset with the world of Thomas Hardy at the turn of the century, and also with the country watercolours of Gordon Beningfield. Unfortunately, in past visits, their Dorset has always remained elusive and limited to small backwaters, off the beaten track, for like the rest of lowland Britain since 1945 Dorset's old landscapes have suffered from the advances made by modern agriculture – although to the casual observer the damage done is not so immediately obvious as in the wheatlands of East Anglia.

My worries were ill-founded; the farm I visited is run economically. But throughout the 1500 acres of the Wrackleford estate the best features of 'old Dorset' remain, giving a rich tapestry of hill and vale, arable field and water meadow, shelter belt and established woodland. The shoot is largely managed by Christopher Pope, a famous name in the area, as his great-grandfather was one of the founders of the local Eldridge Pope brewery.

The farm lies in the valley of the Frome – a beautiful, clear river where salmon and trout breed and moorhens, herons, kingfishers and wildfowl are part of the everyday scene. The river actually flows through the garden of Wrackleford House, and when I arrived a family of recently fledged grey wagtails were feeding on the lawn, down to the water's edge.

The farm, run by Christopher Pope's brother Thomas, is mixed – half devoted to dairy cattle, beef and sheep, and half down to cereals. Much of the grazing land winds with the Frome in the form of lush water-meadows, bringing to life the words of Thomas Hardy;

> I enter a daisy-and-buttercup land,
> And thence thread a jungle of grass:
> Hurdles and stiles scarce visible stand
> Above the lush stems as I pass.

It is a world of hedgerows, meadows and the sounds of water and birdsong. Because of their richness, part of the meadows has been declared an SSSI.

As evening falls and the mist rises it is also one of those increasingly rare places, where white buoyant wings fan silently by, as barn owls quarter the ground. They breed on the farm, for the meadows supply good hunting for mice and voles, and the numerous old buildings and trees give a choice of nesting sites.

Other birds of prey breed on the farm, including buzzards, kestrels, sparrowhawks, hobbys, tawny owls and little owls. Christopher Pope is far from concerned: 'We love to see them. They give us no worries whatsoever. Any damage caused by them is at a very low level.' Recently, too, they had a rare visitor, a black kite. By 6 a.m. the day after its arrival over 100 'twitchers' had turned up. Their lines of communication are good and

Badgers enjoy an abundant and varied diet on Christopher Pope's farm.

they come flocking in from all parts of the country. If the birds are trying to breed or are exhausted, all the attention and disturbance can cause problems, but strangely, some of the conservation bodies seem very reluctant to deal with the harmful aspects of twitching – probably because a number of their own members are involved.

The twitchers who arrived to see the black kite came from as far away as Humberside and Wales. Some had problems with the police who suspected them of being part of an illegal hippy convoy. Wisely the black kite did not stay in the area long and quickly moved on. White storks also visited the farm at the same time; a fact fortunately missed by the twitchers. The latest arrivals have been little egrets – now resident – possibly feathered proof of global warming.

Away from the river there is a mixture of downland and valley. The arable land grows good cereal crops, but some of the farm roadways and hedges around them have thistles, nettles, and a variety of 'weeds', providing shelter and good food for game, wild birds and insect life. The grass valleys are pure 'old Dorset', with woodland planted for shooting by Christopher Pope's great-grandfather along the valley sides. It is all in direct contrast to prairie farming land, examples of which can be seen from the farm boundary.

The woods contain beech, sweet chestnut, field maple, Norwegian maple and the evergreen Holm oak, or Spanish oak – a hardy tree introduced from the Mediterranean over 400 years ago. It provides good shelter from the cold winds blowing in from the Atlantic. In and around the woodland, laurel, privet, hazel, wayfaring trees, shining honeysuckle (a native of China) and even redcurrants provide additional ground cover. In one wood ground cover is also created by 'plushed hazel' – the old method of almost cutting through the hazel and bending it over to grow horizontally.

Planting and re-planting is still going on, as well as work, thinning and clearing, to let light in and to create sunny glades, which not only provide good conditions for nesting pheasants but also encourages the bluebells, wood anemones, early purple orchids – and butterflies. Chiffchaffs, warblers, woodpeckers, jackdaws, rooks and jays

A landscape created by pheasant shooting interests.

all do well in the woodlands and shelter belts, as do the visiting cuckoos. At one time the attractive jays were disliked by many shooting men, because of their egg-stealing tendencies. But now they are often left, as their raucous, scolding call, when disturbed, makes them good watchdogs. On occasions their warnings can be false, simply telling of ramblers, or people on a Sunday afternoon walk, but their screeches can also warn of poachers.

Badgers are other residents of the woods, with plenty of berries, fruit, grubs, worms and roots for a healthy diet, and abundant leaves and grasses to ensure a comfortable underground bed. Roe deer are numerous, and on the increase, while the mixture of woods, grassland and cereals is ideal for hares. Foxes would thrive, if given the chance, but they are controlled; even so Christopher Pope says: 'If they ever became endangered we would leave them alone.'

One steep sided valley with sheep, foxgloves and thistles, together with a beech copse on a hilltop is straight out of Hardy. From the high land it is possible to see the monument near the coast of that other Dorset Hardy, Vice-Admiral Sir Thomas Hardy, of 'Kiss me Hardy' fame. Adjoining the valley is an 'ancient monument' – 200 acres of old grassland over the site of a neolithic village, a haven for skylarks, corn buntings and meadow pipits. Christopher Pope is not an expert on butterflies, but he likes to see them and confirms the presence of 'blue butterflies' in the summer. One of the shoot beaters, however, is an expert and has identified the common blue, the chalkhill blue, the brown argus, the large, small and dingy skippers and the rare marsh fritillary. Another fritillary, the silver-washed, is to be found in the woodland. In 2002 Butterfly Conservation found a flourishing colony of the marsh fritillary in one of the low lying meadows. The farm is rich in birds, wildflowers, insects and animals; all flourishing in a landscape that has been both created and preserved through shooting.

This pattern of coppice, shelter belt, grassland, game and wildlife is not limited to large farms and estates, as it can be found on some of the smallest holdings. At one end of the scale I know of a landowner in Norfolk with just sixteen acres of old grassland and sprawling hedges – perfect for cowslips and blackberries, as well as the one hundred pheasants he releases each year. He holds several shoots during the course of a season, with four beaters and five guns. Each shoot is over shortly after mid-day, when the guns retire to a meal and a log fire. The 'bags' (the number of birds shot at the end of each day) are small, but those taking part enjoy the sounds and scents of the winter landscape, and the small fields form a small wildlife island in a sea of East Anglian prairie.

At the other end of the scale the Duchy of Cornwall also encourages shooting, specifically as a means of improving the landscape and creating good wildlife habitat. It is a way of stimulating interest in the whole spectrum of natural history, from the trees to be planted, or left in the hedgerows, to the insects and grubs favoured by pheasant chicks. On the Duchy's Bradninch estate in east Devon, fifteen tenants have joined together in a 'tenants' shoot', with encouragement from the local Land Steward.

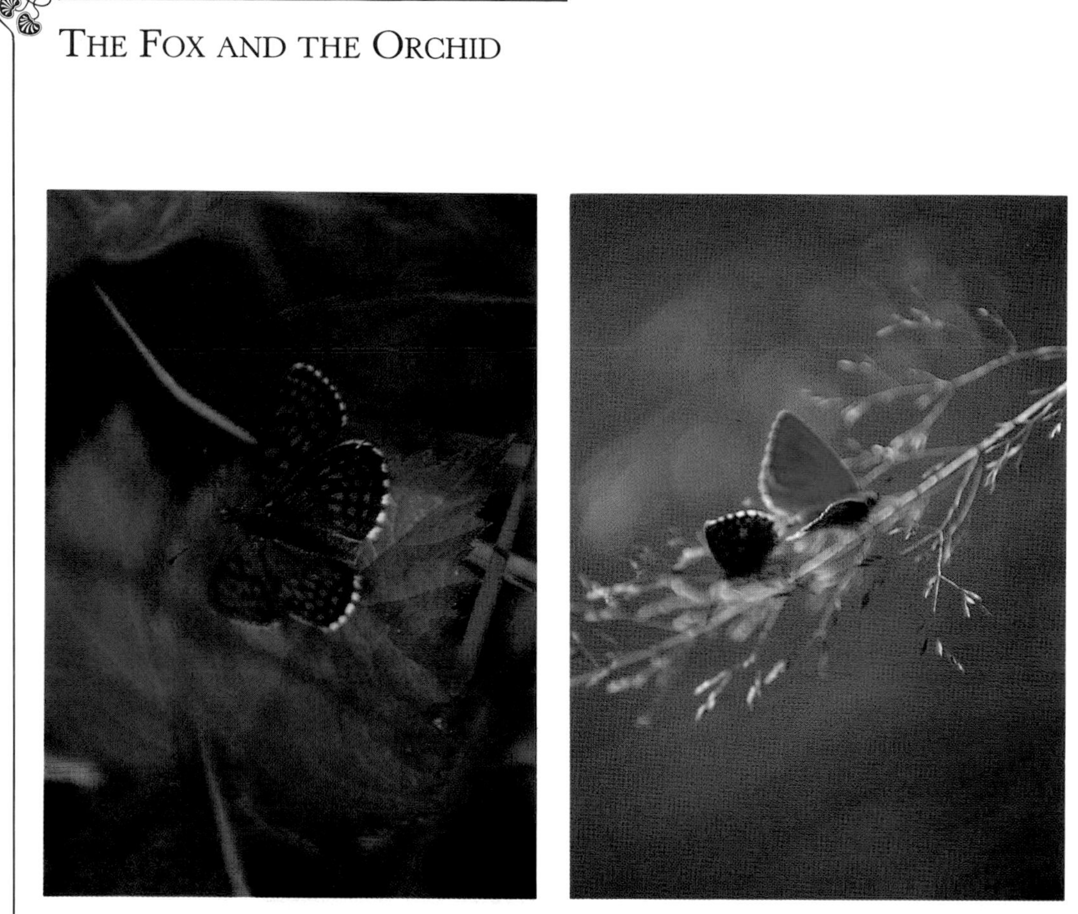

The Marsh fritillary (left) and the Common Blue.

Although seven of the fifteen don't actually shoot, it is very popular and they are all interested in the work being done. It helps with a feeling of community, too, and each year well over seventy people get together to enjoy the shoot supper. Tree planting is being encouraged, wetland created, and trees are being left to grow. Indeed some hedgerows have been 'let go' completely, which means that tenancy agreements are being re-drafted to prevent wild hedges from being classified as 'dilapidation'. This development is one that ought to be copied by other landowners with tenant farmers – whether the landowners be private individuals, or institutions. Because of the co-operation between the Duchy and its tenants, to create a better countryside, the estate stands out from some of the neighbouring land where there is no game interest; there the dominating influence on the landscape is straightforward commercial farming.

All over the country the story is the same. Because of the pheasant, many thousands of acres have been spared from the plough, the chemical spray, the chain saw and the flail mower. Through the pheasant we have an unexpected link with our historical past, and it is through the pheasant that we can help save what is left of our traditional countryside for the future.

Chapter 6

WHERE PARTRIDGES STILL FLY

Like the story of the hare, the recent history of the grey, or English partridge is depressing; it is one of falling numbers and continuing decline, causing some people to forecast its eventual extinction. At the turn of the century it was one of the most common birds on lowland farmland; yet today there are huge areas of highly intensive arable agriculture where the partridge has become a rarity. In fact since the 1950s the spring population of the partridge has dropped by 93% and it is still declining at the rate of 7% a year, and shooting bags have fallen from 2 million per year to 100,000. A number that is only maintained by adding reared birds to the wild stock. To make matters worse chick survival has fallen from 50% in the 1930s to 25% today and it is considered that another drop of 5% could lead to the bird's total disappearance.*

Unfortunately modern farming methods have allowed no room for the partridge to live or breed, and the chemicals used to stimulate plant growth and to protect crops have reduced the most vital elements in a young chick's diet, making survival less likely. Even where cereal crops are not grown and grass predominates, the story is the same, for grass can be cut several times each summer, with forage harvesters destroying not only the nests of the English partridge, but also killing the hens themselves; for they sit tightly on their eggs even when danger threatens. Sadly in many parts of England the 'meadow' has become the 'grass production unit' and the 'farmer' a mere 'producer'.

Again the change has been rapid. Before the last great surge of agricultural advance, the English partridge flourished so well that one of its everyday country names was the 'common partridge'; that name is now totally inappropriate – a dubious distinction it shares with the unfortunate 'common frog'.

Although many people call it the grey partridge, and some scientists refer to it as *Perdix perdix*, I have always known it as the English partridge. It seems to be the most appropriate name as the bird is indigenous to Britain, unlike the French, or red-legged partridge, which was introduced from the continent. The first recorded 'Frenchman' dates back to the seventeenth century, with most of the early introductions taking place the following century. They are strikingly beautiful birds and their range goes far beyond France. They have now been living wild in Britain for so long that they can be regarded as native; but because they are easy to rear and considered to be only moderate sporting birds, the French partridge will never be regarded with the same affection as the English, either by the naturalist or the shooting man.

From very early times the partridge was highly regarded as a bird for the table. Before the development of the sporting gun it was netted and snared, sometimes with the help of setters, pointers or spaniels.

The use of snare or net also helps to explain a line by John Gay (1685-1732) in *The*

---

* See *The Partridge* published in 1986 by Collins for more details. It is an excellent book written by Dr G. R. Potts, The Game Conservancy's former Chief Executive.

Partridge shooting – an engraving by Sam Howitt 1807.

Beggar's Opera: 'The good sportsman always lets the hen partridge fly, for on them depends the breed of game.' With a gun it is not possible to distinguish cock birds from hens, in flight. But it would be possible to release any hens caught up in snares or nets.

James Thomson in 'Autum' also mentions netting as a once common method of taking partridges:

> As in the sun the circling covey* bask
> Their varied plumes, and watchful every way,
> Through the rough stubble turn the secret eye.
> Caught in the meshy snare, in vain they beat
> Their idle wings, entangled more and more.

Falconry was another popular country sport in which the partridge was often the quarry.

With the advent of the sporting shotgun, partridge shooting's popularity quickly spread. The birds' fast flight and the reactions required to shoot them made it a

---

*A family of partridges

ABOVE LEFT: *Partridge chicks under a bantam hen.* ABOVE RIGHT: *The English partridge – now returning to some farms, with help from the farmer.*

challenging sport. At that time partridges were more numerous than pheasants and partridge shooting became the most common form of game shooting. In addition partridges make an excellent meal – almost as good as pheasant.

When I was a boy, partridges regularly appeared on our dinner-table during the winter and the sight of coveys feeding on the autumn stubble or planing low over a hedgerow were part of the everyday parish scene. Whenever a hen bird was accidentally cut off the nest at hay time, Jim would collect the eggs in his cap and set them under a broody hen or bantam. After the tiny chicks hatched we would dig into ant hills to get the white 'eggs' (pupae) as the young birds were said to thrive on them. It was a method of rearing game chicks learnt many years before and practised by succeeding generations of countrymen. In the nineteenth century, Thomas Bewick, the outstanding wood engraver, wrote: 'It is no unusual thing to introduce partridges' eggs under the common hen, who hatches and rears them as her own: in this case the young birds require to be fed with ants' eggs, which are their favourite food, and without which it is almost impossible to bring them up. They likewise eat insects, and when full grown, all kinds of grain and young plants.' Today some university-trained ecologists, biologists and various other high-faluting '-ologists' tend to look down on the straightforward observations of country people as 'unscientific'. Yet, over recent years, scientists at the Game Conservancy have confirmed the observations of Thomas Bewick - that young partridges need insects in their diet. The scientists have now gone a stage further by showing that it is the loss of much insect life in the countryside that has led to a corresponding loss in the partridge population. Natural history is observation; 'science' is simply recorded observation, measured and analysed.

Partridges thrive at the weedy edges of corn fields where sprays have not killed the wild flowers and the insects that depend on them. A well known Rodger McPhail painting.

Thomas Bewick was a very good observer, as his woodcuts in this book show, and he was obviously fond of the partridge:

> The affection of the partridge for her young is peculiarly strong and lively; she is greatly assisted in the care of rearing them by her mate; they lead them out in common, call them together, gather for them their proper food, and assist in finding it by scratching the ground; they frequently sit close by each other, covering the chickens with their wings, like the Hen. In this situation they are not easily flushed; the sportsman, who is attentive to the preservation of his game, will carefully avoid giving any disturbance to a scene so truly interesting.

We still have partridges on the farm, both French and English. With our hedges, rough ditch sides and unkempt brook banks the insect population is high; it is probably why we have only had to spray for aphid infestation of crops on two occasions. The rough brook banks are of value to more than insects, for in severe winter weather the partridges make direct use of it for shelter. With the arrival of the CRT's Lark Rise Farm next door the amount of partridge friendly land has nearly quadrupled and the English partridge is again common.

It gives me great pleasure to see a covey of 'Englishmen' running low over the plough, or bursting into flight after being surprised sunning themselves. Their distinctive grating call also makes a soothing, pastoral end to a summer evening — something enjoyed by John Clare too: 'The partridge has a very pleasant call in the evening among the wheat calling its mate or young together.'

It is significant that the link between partridge decline and a loss of insect numbers was established by the Game Conservancy as part of its monitoring of farm practices – particularly the effect of farm sprays on game and wildlife. The Game Conservancy has carried out this watch-dog function on the countryside for many years (formerly as the

Eley Game Advisory Service and later as the Game Research Association) and it was largely its work on dieldrin that alerted the other agencies and organisations to the grave threat posed to birds of prey and otters during the 'sixties.

Since then the Game Conservancy has undertaken important work to show how game birds can co-exist with modem farming in its 'Cereals and Gamebirds Research Project'. Many of the experiments relating to partridges took place on the farm of Hugh Oliver-Bellasis, in Hampshire, which he runs with his brother. The farm itself has been held by the family for over 140 years. It is 5000 acres in size, 3000 of which is farmed 'in hand'. There are 380 acres of woodland including 120 acres of oak and hazel coppice with its associated wood anenomes and bluebells. Of the 'in hand' land over 1500 acres are usually put down to cereals; rape, peas, beans, potatoes and linseed are also grown, and there are about 140 acres of old parkland and permanent pasture.

Before 1939 the estate was noted for its partridge shooting and over 130 brace were shot every year. Numbers fell during the Second World War as no 'keepers were employed and the population continued to fall afterwards, until an all time low was reached in 1965, which extended on into the 'seventies, with just a brief respite during the drought year of 1976. The population probably dropped to as low as sixty pairs for the whole estate.

Because of this, Hugh Oliver-Bellasis was so worried in 1982 that he decided to leave the headlands of eight fields unsprayed, to try and improve numbers, as he was very fond of both partridges and partridge shooting. There was an immediate and obvious improvement in chick survival and the population has been increasing ever since; so much so that it is now possible to hold three partridge shoots a year.

In 1986, 40% of the fields had unsprayed headlands but since 1987 all the headlands of cereal fields have been unsprayed. 'Beetle banks' – raised, grassy strips, ideal for birds and insects, cross every field and as a result partridge chick survival has more than doubled and the numbers of insects, butterflies, wildflowers and arable 'weeds' have all increased significantly. Hugh is not the only one who is pleased, as his old gamekeeper always maintained that: 'If the partridge is in good health – so the rest of the farm will flourish.' He was perceptive; today the scientists mean the same thing when they call the partridge an 'indicator species'. If the habitat is changed or harmed then the partridge is one of the first species to decline, 'indicating' a new, harmful factor, or factors, in the environment; other important and sensitive indicator species are the hare, butterflies and, on moorland, the grouse.

Since the initial work, Hugh has changed the phrase 'unsprayed headlands', to 'conservation headlands', for greater accuracy. He does spray herbicide in the autumn, to control harmful grasses and wild oats, but he will not spray either herbicides or insecticides in the spring, just fungicide, as that has been shown to be harmless (non-toxic) to broad-leaved weeds and insects. The aim is to get a good population of broad-leaved weeds, for they encourage high insect populations, which feed on them, and in turn they

The unsprayed headland with its flowery bonus.

The rare pheasant's eye and the prickly or pale poppy, both on the estate of Hugh Oliver-Bellasis.

help the survival of the partridge chicks. The headland is the most important part of the field, for it is the arable area adjacent to the hedge bottoms and rough grassy strips where partridges nest and hatch their young. The insects most favoured by the chicks are various plant bugs, weavils, leaf beetles, rove beetles and the caterpillars of sawflies, butterflies and moths.

Hugh admits that there is some loss of quantity and quality in the cereal yield of a conservation headland. To overcome this he feeds the barley to his beef cattle and the wheat to his reared pheasants (approximately one pheasant for every two acres). He is realistic about his experiments, and because of his background and large farm he describes himself as 'a silver spoon job'; as a result he can see that small farmers could not afford to do what he is doing, without government support. Even so he reluctantly concedes that despite all the conservation plusses, even he has had to gradually intensify his farming methods because of the pressure of world markets – falling farm gate prices on his farm income and there can be no farmland conservation without profit – something that seems to have escaped the notice of our politicians.

The results of all the conservation work so far are extremely encouraging and have important implications for the future of farming. Although some of his arable land had been sprayed annually for the previous twenty years, numerous 'weeds' returned in the first year without spray. When the weed populations were studied, of the twenty-five species said to be in decline by The Botanical Society of the British Isles, fifteen had reappeared in one field, due to the fact that arable weed seeds appear to be able to remain dormant in the soil for many years. Because of this Hugh prefers not to call them 'endangered', but 'herbicide suppressed'. Among the now scarce arable weeds or, more accurately, wildflowers to return are the prickly-headed poppy, long-headed prickly poppy, field gromwell, shepherd's needle, venus' looking glass, corn salad, dwarf spurge and pheasant's eye. The names of the plants are as attractive as the flowers themselves, reminding of the words of H. E. Bates in his beautiful book, *In the Heart of the Country* published in 1942. He wrote, 'But some day someone will write that history [of weeds]. Someone will show the part played by weeds in the lives of animals and birds, and so in the life of man, and how, both economically and aesthetically, life would be poorer, perhaps even dislocated, without them.'

The increase in wildflowers has not only led to an increase in various bugs and beetles, but it has been of great benefit to butterflies. Twenty-four species have been found on the estate in recent years, including butterflies of woodland, hedgerow and meadow.

Like John Wilson in Suffolk, the desire of Hugh Oliver-Bellasis to improve his wild-game stock has developed into something much wider. This was shown to me as I rode around the estate with him, for he was enthusiastic about many forms of wildlife. He stopped to show me a small ridge, about 350 yards long. It was deliberately created through ploughing, to provide an ideal site for a new hedge. I was puzzled as no quickthorn or young hedgerow trees had been planted. Hugh explained: 'We are leaving the ridge for three years before planting, to allow the hedgerow weeds to come back, and then we will plant the quickthorn. The ridge is to provide a well drained site for the young trees – after all you wouldn't put your bed in a gutter, would you? Yet the Ministry of Agriculture just says spray and stick the quickthorn in.' The hedge will be an addition to the farm, for very few hedges and no woods have been lost since the Second World War. The hedges on tenanted land have also been retained and because of the specific hedge management required, together with rough bottoms, an allowance is made in the rent charged.

His field edges and the sides of his farm roads and driftways are left rough, with thistles and nettles, as well as banks of field scabious and knapweed. 'All these things are good for wildlife,' he observed. 'Indeed I think that in "good farming" competitions, marks should be deducted from people who spray or flail all their nettles and thistles and odd corners. We know now that it is not "good farming" – but "tidy farming", and it can be very harmful.'

In fact thistles and nettles are the food plants of a surprising number of insects and beetles, most of which cause no threat to farm crops whatsoever. The most well known feeders on nettle leaves are the black caterpillars of the small tortoiseshell and peacock butterflies. In turn many species of adult butterfly are fond of thistle flowers, for their nectar.

Because of the way in which the farm is run, it still has at least two pairs of resident barn owls and many hares (with the accompanying problem of didecoys and lurchers). The mixture of field, hedgerow and woodland is also attractive to roe deer, and they have recently colonised the area, moving in from the south-east. Finally, Hugh showed me a patch of woodland which gave him no pride at all. It was 4½ acres of conifers and beech, planted to the Forestry Commission's standards under the Small Woods Scheme. He describes it as an ecological disaster area – being dark, with bare ground and pine needles beneath. For wildlife it is virtually useless, while for pheasants it offers a small amount of shelter in winter; he will never plant another similar block. The pioneering innovations of Hugh Oliver-Bellasis and The Game Conservancy have been of immense environmental importance. The CRT quite openly copied and modified the successful methods used by Hugh and they are now commonly used in wildlife friendly farming; grass margins, beetle banks, conservation headlands, wildlife strips etc are all the result of this work. Rather belatedly other bodies such as the RSPB have become involved with practical farming issues – but seem somewhat reluctant to acknowledge the effort of the real pioneers. There is no doubt that the work done by Hugh Oliver-Bellasis and The Game Conservancy forms the basis of the various agri-environment schemes gradually being implemented across Europe.

But although the estate shows how good farming and wildlife can co-exist, it also reveals another growing problem that can be found throughout Britain. Sprays, fertilisers, flail mowers and over-intensive farming are not the only threats to the general countryside and its wildlife, for there are numerous difficulties caused by access and people. The land farmed by Hugh and his brother lies close to Basingstoke – the commuters' paradise, and the 'village' of Oakley.

When Hugh was a boy Basingstoke had a population of 17,500 – in 1987 it was 130,000, now he tells me, it has grown to 155,000, in sixteen years. During the same period Oakley has grown from just 345 inhabitants to 7500 at the time of the first edition of *TheFox and the Orchid* and to 8500 in 2003. The 5000 acres of the estate have twenty-six miles of public footpaths, used by as many as 50,000 people a year. Unfortunately, at crucial times during the breeding season of wild birds (including game birds), many walkers do not keep to the paths. Consequently of all the partridge nests found, 26% are lost to predators and of that 16% are lost to 'people pressure'. Dogs are allowed to run loose, eating eggs or chicks, or simply causing the sitting birds to desert. Some people accidentally tread in nests while walking off the beaten track, and others deliberately steal eggs.

In the early 'eighties Hugh spent £1800 signposting the footpaths; at the time of my visit only six were left out of a total of 340. The local Chairman of the Ramblers' Association explained the damage by saying that the signs were the wrong colour – red, instead of green. So Hugh erected 120 new green ones; 117 have since disappeared and one was even removed with the aid of a blowlamp. The mind boggles – what sort of strange individual walks through the countryside carrying a blowlamp? One day he saw a couple wandering off the footpaths towards a field of rape that was in the process of being sprayed. He stopped to warn them of the spray, and was greeted with a tirade of abuse about wealth, landowners and how people had a right to go wherever they liked.

Some of the footpaths are used as motorcycle tracks and the whole farm serves as an urban litter bin. Each year it receives an assortment of old refrigerators, microwave ovens, 'Christmas jollity waste', road rubble, bedsteads, motorbike frames, garden rubbish and used contraceptives. In addition, about £10,000 worth of damage is done annually, with crops flattened, fences broken, straw burnt and wood stolen. In recent years there have been eight major fires at harvest time; one burning between twenty and thirty acres of straw and running with the wind towards Basingstoke. Because of this an old fire-engine has been purchased. Straw and hay bales are damaged every summer and those near houses have to be baled and carted on the same day; in 1980 a whole barn containing 100 tons of straw was burnt down and finally, just to round things off, newly planted trees are often uprooted and taken away to private gardens. In recent years fly-tipping, car dumping and illegal coursing have become out of control and some neighbours shoot all their hares to keep illegal coursers off their land and to give themselves peace of mind. Farm buildings are all kept locked – but the locks are broken regularly and from one barn an all 'terrain vehicle' – an ATV – was stolen, despite being immobilised and screwed into concrete. The police seem unable to cope with the tidal wave of crime and as a result sharply rising farm insurances create another financial problem.

Although Hugh Oliver-Bellasis has led the way in restoring partridge numbers, using techniques being followed by an increasing number of farmers, there are a few farms where the populations of the English partridge have remained consistently high. One such farm can be found on the South Downs close to the River Adur, farmed by Christopher Passmore. His grandfather moved to the holding in 1901, but signs of arable occupation go back many years, for there are clear indications of ancient Celtic fields, dating back to the Early Iron Age – some time between 1200 and 1 BC.

It is thought that the Downs were once covered by beechwoods but clearing began in the Stone Age. In more recent times the Downs have been known over many generations for their grassland and sheep. Much was ploughed during the Napoleonic Wars, when the price of corn was high. The sheaves were taken from the hill tops on sledges to the waiting wagons in the valleys below. Then, when corn prices fell, the area again reverted to grass and sheep.

In the middle of the nineteenth century there was another spell of arable production, and the remains of old barns and yards can still be seen dotted around the farm, where corn was stored and horses and oxen were housed – to save walking large distances at the beginning and end of each day, to the main farm buildings. One old local man could remember the oxen era and told Christopher Passmore that they would be worked eight

Red campion flourishing in a rough edge of Christopher Passmore's down-land farm.

to a team. They were cheaper to keep than horses, but could only be worked two or three days a week. Their other main disadvantage was their two speeds – 'dead slow and stop'. Because horses were stronger and faster, the oxen were gradually replaced. The yield for cereals in those days was about half a ton an acre – today if a farmer does not get three tons an acre he is disappointed.

During, and after, the Second World War ploughing of grassland again became fashionable, to increase cereal production. It should be remembered that this ploughing was the policy of successive governments to increase Britain's self-sufficiency in farm produce. For many years a grant was paid as an inducement; consequently farmers should not receive all the blame for today's food mountains and loss of landscape. Those most directly responsible are the politicians – very often the same ones who now complain of surplus production. Of course, to pay a grant for the production of something that is not wanted in the long term is rank stupidity; strangely, in the never ending stream of words that flows from our politicians, I have never heard them criticise their own inept performance or lack of foresight.

Applesham Farm clearly shows its history; it has smooth, rounded, treeless hills, and large open fields. Some of the old field names record its past – Sheepwash, Horsebrook, Top Down, Cow Bottom, Forty Acres, Hill Barn and Ladywells. Today, Christopher Passmore produces wheat and barley, but he also has beef cattle and sheep, for which he

A South Downs dew-pond preserved on the Passmore farm.

grows forage rape and rye grass. In addition he 'undersows' some cereal crops with a mixture of grass and clover. This means that cereals and grass are sown together; then, after the combine harvester has cut the wheat or barley, the field is left for grass and grazing. It is a form of husbandry based on crop rotation, with each field growing cereals for four or five years, and then grass and clover for three. Once more it means an abundance of insects, which in turn has resulted in the population of the English partridge remaining healthy. Christopher Passmore gets excited when he talks about insects: 'It's incredible when you think of it,' he says. 'It's like a miniature Serengeti down there – things like aphids and their predators – the grazers and the scoffers. I hardly ever have to spray for aphids; I think it is crazy to spend a pound on spray to make a pound.'

Despite the absence of hedges, the partridges nest in patches of rough grass and nettles along the fence lines, and because of the high insect population chick survival is good, and partridge shoots are held every year. Local country lore does not always agree, as it is said, that, 'If it rains during Ascot week (third in June) there will be no partridges.' There is probably some truth in this, as June is a key time for partridge chicks and rain would be bad for them and the insects they feed on. If chicks have to search far for insects in the wet, they will die of chill: sometimes they like to dash out for a snack and go back to mum for brooding.

Among the grass fields there are seventy acres of old unimproved chalk grassland, which is grazed in rotation. It is rich in plants, butterflies and a variety of insects. One summer day Christopher Passmore went on a short walk to count wildflowers (excluding grasses); he expected to get about forty, but he totalled over one hundred very quickly. He has three types of orchid on the farm; the pyramidal, spotted and bee, as well as cowslip, harebell, salad burnet, horseshoe vetch, hop trefoil, dwarf thistle, welted thistle, 'nodding' thistle, red bartsia, fairy flax, hounds tongue, wild carrot, round headed campion – also known as 'the pride of Sussex' – and many more.

It is more than a botanist's paradise, for it is full of ant hills, grasshoppers, and countless creatures that hop, crawl and fly. The butterflies have not been surveyed, but they include 'several blue varieties', 'common brown ones' and marbled whites. From the grassy hilltop, surrounded by sunlight and wildflowers, there is a beautiful view. Recently the whole area was pronounced an Environmentally Sensitive Area, to ensure that it is not destroyed. On Christopher Passmore's land this was not necessary, but on some other downland in the district nearly all the old flora and fauna would eventually disappear without statutory protection.

The farm has many skylarks, meadow pipits, lapwings (or 'peewits') and corn buntings, birds that have disappeared from some nearby land which is more intensively farmed, and every spring wheatears move through. In the winter, short-eared owls arrive to hunt over the old grassland. Despite the sensitive farming methods, hare numbers have declined to about a quarter of their former population. It could be that once they leave Applesham Farm they have more sprays and machinery to cope with. Because of the hares there is again a didecoy and lurcher problem, made worse by magistrates who do not seem to understand the country. On one occasion poachers were observed walking up one side of a hill; with the help of two-way radios they were caught by the police as they went down the other side. The case was dismissed by the magistrates because of 'uncollaborated evidence'.

In view of the decline in hare numbers, Christopher Passmore no longer allows the local beagle pack to hunt his land. The fox hunt is welcomed, however. The farm is only lightly keepered, for foxes, crows and the occasional magpie, and so the hunt helps with the removal of a few foxes. Most years several lambs are lost: 'Some foxes get the knack – not all foxes. They only take very young lambs. If you lamb indoors, which we don't, that would almost elminate the problem, as you move the lambs out when they are bigger. We make a lambing ground out of big bales and put an electric fence round it; foxes don't like that, but they still take some. The losses are not heavy, but they do happen. Two years ago we were losing one or two lambs every day for a week or ten days. It's nearly always one of a twin; a mother will defend one but finds it difficult with two.' He also believes that dead lambs should be taken away and buried:

Partridge shooting – and retrieving by the ubiquitous English spaniel.

'If they start off on dead lambs, then they get the taste.'

A footpath runs through the farm, but on the whole he gets little trouble from walkers. His main problem is from a nearby County Council refuse tip, where plastic bags and papers blow onto his land. At one time it was so bad that he went to the County Council Offices to obtain some 'Keep Sussex Tidy' leaflets; he left them a few doors along the corridor in the Engineer's department.

Visiting Applesham Farm was my first real view of the South Downs. In addition to the old landscape there were some newly planted woodlands, to replace the losses caused by Dutch Elm Disease; three ponds, including a restored 'dew pond', and fifteen acres of traditional marshland grazing. But my strongest memories are of the old grassland, with grasshoppers chirring, and skylarks singing; the fields of grass and clover, with contented sheep and cattle, and a field of autumn stubble with three coveys of English partridges. One covey contained eighteen birds, with the sun clearly catching the dark horseshoe outline on their breasts. Over recent years the English partridge has suffered considerably; but with help and care it is on the way back.

Chapter 7

GROUSE IN THE HEATHER

The most well known date in the country sports calendar is August 12, the first day of the grouse shooting season. It is also the time that attracts the greatest amount of ridicule. As the day dawns, the media overflows with stories of the 'upper classes blasting defenceless grouse out of the skies'; the favourite media word of the day is 'slaughter'; ramblers complain about lack of access to grouse moors, and almost ritual-istically, grouse 'butts' (the small, constructed shelters behind which grouse-shooters shoot) are knocked down including some in the Derbyshire Peak District. The fact that damage is done in Derbyshire is significant, as the Peak District and its grouse moors lie between Manchester and Sheffield. Both these sprawling connurbations contain various 'animal activist' groups, as well as 'radical ramblers'. Living so close to an attractive piece of rural England it could reasonably be expected that they would try to understand the wildlife issues on their doorstep. Sadly, their knowledge of the moors and the impor-tance of grouse to a host of flora and fauna seems to be beyond them.

Those at the other end of the idiocy ladder do not help matters either, for each year on the twelfth, grouse are whisked from the moors, in planes, trains and helicopters by commercial interests, each one trying to get the birds to London, so that they can be on the lunch-time menu. Somebody should tell them that grouse need to be hung for at least a week before being roasted. The birds should then be eaten with sausages, bread sauce, parsley and thyme stuffing and red currant jelly or cranberry sauce. As it is rather a dry meat, it is better if cooked on pieces of buttered toast, with strips of fat bacon over the breast. If hung and cooked properly it makes an excellent meal.

The importance of grouse to our highland wildlife heritage cannot be over-stated.

The red grouse – important in the preservation of our heather moors.

Just as the pheasant, partridge and fox have prevented much of lowland England from falling to the worst excesses of the barley barons, so grouse have played an even more important role in preventing afforestation and over-grazing by sheep, over vast tracts of heather moorland. Huge, wild areas of Northern England, Wales and Scotland owe their survival to grouse; without them the irate animal activists and radical ramblers would have far fewer acres of moorland over which to walk and feel aggrieved.

The grouse is an attractive, secretive bird, but when disturbed it will burst into flight, calling raucously as it goes. For years 'experts' claimed that it was a bird unique to Britain. Now they have changed their minds and decided that it is a variety of willowgrouse; one that does not turn white in winter, like its close relative the ptarmigan. It may not turn white, but it does have well feathered feet to help it get through the worst of the cold weather. It lives mainly on heather, favouring ling, which forms 90% of its diet. Its dark mottled plumage makes excellent camouflage. A heather moor in the spring, with curlews calling and the cock grouse displaying, is a wonderful place to be. In its efforts to impress the female the cock bird flies straight up – simply to glide down again calling loudly. According to James Grahame (1763–1811) courtship coincides with the heather beginning to bud:

> With earliest spring, while yet in mountain cleughes (gorges)
> Lingers the frozen wreath, when yeanling (yearling) lambs,
> Upon the little heath-encircled patch
> Of smoothest sward, totter, the gorcock's (red grouse) call
> If heard from out the mist, high on the hill;
> But not till when the tiny heather bud
> Appears are struck the spring-time leagues of love.

An extensive heather moor attracts far more than grouse, however; other breeding birds include the merlin, dunlin, curlew, golden plover, redshank, raven, wheatear, meadow pipit, peregrine, short-eared owl, hen harrier, ring ouzel and, in Scotland, the golden eagle. Britain's most attractive moth, the emperor moth, is another inhabitant, together with the dark green fritillary butterfly in some areas. Heather is not the only plant either, as there is bearberry, crowberry, cowberry, cloudberry, cotton grass, bog asphodel and many more, as well as specialist plants of the arctic tundra in the higher and more northerly regions. In the heather and 'lochen' country of Caithness and Sutherland, other birds also find the habitat important – in fact the 'flow country' as it is known, accounts for 81% of Britain's breeding greenshanks, 11% dunlins, 21% golden plovers, 41% black-throated divers and 95% common scoters.

As a result a grouse moor is far more than a place where shooting occurs amid much ballyhoo on the 'glorious twelfth'; it is a complete 'eco-system'. Without moorland

ABOVE: *The merlin – one of our most endangered birds of prey but one which can breed in captivity.*

RIGHT: *The golden eagle – the symbol of wilderness.*

being specifically managed for grouse, the heather declines and the whole range of wildlife associated with it slowly disappears. The importance therefore of the red grouse is acknowledged by many conservation bodies including English Nature, Scottish Natural Heritage, local wildlife trusts and the RSPB. To confirm this further, some National Parks actually encourage heather management for grouse shooting. In 1984 Sir Derek Barber, Chairman of what is now the Countryside Agency, said: 'From numerous and varied sources comes irrefutable evidence, scientific, socio-economic and the rest that, but for land owners nurturing the grouse interest, heather moorland, which is an increasingly highly-valued national asset, would have all but disappeared in some uplands. The argument is indisputable.' It was indisputable then and it is indisputable now.

Wisely, the RSPB is prevented by its charter from taking part 'in the question of killing game birds and legitimate sport of that character' (which includes falconry). The RSPB's spokesman says: 'The morality of shooting is something that does not concern us and we take a strictly neutral line – however, where there is a conservation issue or principle involved, then we do get involved. Undoubtedly there are many lowland woods and upland moors which owe their existence to shooting practices. Many areas that are not shot over have fewer woods and fewer moors and have become less important for birds. This is not supporting shooting – but saying what has resulted because of shooting, and we recognise that. There is little doubt that if it wasn't for grouse shooting we wouldn't have the upland moors that we have today.' Despite many leading naturalists and conservation organisations confirming the importance of grouse, the ridicule persists, and some 'anti' organisations openly state that they want legislation to ban grouse shooting – not only on the grounds of 'cruelty', but to 'save wildlife'; the ignorance, or hypocrisy involved is almost beyond belief.

Sadly there are far more pressing threats to grouse than legislation, for in many areas grouse populations are falling. Heather is disappearing at an alarming rate because of afforestation and over-grazing by sheep and deer, and there are problems and damage caused by too many people, problems made worse by 'the right to roam'. This mythical 'right' to roam on moorland throughout Britain is loosely based on Sweden's 'Everyman's Right'. The difference of course being the Sweden's population density is just 20 people per square kilometre, Britain's is 241, with England's reaching a staggering 377, higher than the density of 'over populated' India (320 people per square kilometre) and China (132)

If a grouse population is high; it means that grouse shooting can provide a reasonable income from what in agricultural terms is very marginal land. If the grouse declines other forms of income have to be obtained. When sheep numbers are increased the heather dies through over-grazing or mis-management and the grouse never return: nor do the emperor moth or merlin.

Rodger McPhail – Grouse on the wing.

Grouse numbers have always experienced great variations – population explosions and crashes. Unfortunately, over recent years the grouse population of most of Scotland has been in a state of almost permanent crash. To understand the reasons the Institute of Terrestial Ecology and the Game Conservancy have studied the problems for a number of years. The Game Conservancy first successfully ran the North of England Grouse Research Project, and the Scottish Grouse Research Project under Dr Peter Hudson. His main areas of interest were the three 'Ps' – Protein, Parasites and Predators, which all involved aspects of habitat management. That work has continued as English Uplands Research and Scottish Uplands Research under Dr David Baines and a twenty-first century Adam Smith.

From the various studies it is clear that grouse need a variety of heather conditions to survive well, from young shoots for nourishment, and old heather to provide cover and concealment. To help with this variation a grouse moor should be burnt in rotation, giving the familiar patchwork pattern of young and old heather that signifies a working and well-run moor. In addition there should be plenty of boggy flushes where the young chicks can get additional natural protein from insect life. Sheep and deer numbers must be strictly limited to prevent over-grazing.

Parasites play an important part in reducing grouse numbers. The birds are particularly susceptible to the Strongyle Threadworm, and bad infestations can lead to death and poor chick survival. Louping-ill is the other major problem; a virus carried by the sheep tick. The fewer the sheep, and the healthier (with regular dipping against ticks), then the better it will be for the grouse. Although ticks can be picked up in heather, as anybody who has had a snooze on a heather moor will know, they favour bracken. Bracken control is therefore an aid to tick control, and it also provides more grazing as sheep cannot eat bracken.

The final problem concerning grouse comes from predators, particularly the fox and the crow. Birds of prey take a lesser toll, although the hen harrier can cause difficulties.

For moor owners where grouse numbers have declined, and there are clear signs of overgrazing and heather loss, there is hope. When Lord Peel, President of the Game Conservancy, inherited his estate in Swaledale, North Yorkshire, he needed grouse shooting to help pay off the death duties. But the grouse moors were in a poor state; the grouse were declining, the moors were being over-grazed by sheep, and he estimates that in the twenty-five years before he took over, more than 5000 acres of heather disappeared.

As Lord of the Manor he had shooting rights over 32,000 acres, but much of the land also had commoners' rights, which allowed local farmers to graze sheep. Because of the Hill Livestock Compensatory Allowance (a hill sheep subsidy, paid annually on a headage basis for ewes), some farmers were grazing more than they were allowed, to increase their income, and the Ministry of Agriculture was turning a blind eye. To control the grazing and get it back to sensible levels the ancient commons committees

Golden plover, whose nesting sites are disappearing through afforestation.

were revived, and to prevent damage to heather in sensitive areas, sheep were removed from them altogether during the winter.

Lord Peel was not universally popular for what he was trying to do. But with sensible husbandry new young heather returned over large areas. With the improvement and the increase of heather moorland, so grouse numbers increased significantly. When the heather was firmly re-established then sheep were allowed back in controlled numbers. The work done by Lord Peel was as important for heather moorlands as the work done by Hugh Oliver-Bellasis, for cereal land.

Swaledale is a most beautiful part of the country, and I could not understand why I had not visited the area years earlier. The valley is full of small grass meadows, with stone barns and stone wall boundaries. There are small villages, streams and waterfalls, the bleating of sheep and higher up, the grassland gives way to heather. Walking through the heather with Lord Peel was an unusual experience; he kept stopping to look at its quality, rather like an East Anglian arable farmer inspecting his wheat. The similarity was not strange, for heather and grouse were his crops, and shooting parties, particularly from America, paid for the privilege of staying with a real live lord and shooting his grouse.

The finances of grouse were not Lord Peel's only concern. He is the great-great-grandson of Sir Robert Peel, and by coincidence his grandfather once owned The Lodge, at Sandy in Bedfordshire, which is now the headquarters of the RSPB. Appropriately Lord Peel was interested in all the fauna and flora of this moorland, including the birds.

The goshawk after killing a red grouse. It is a most efficient killer.

His land was rich in breeding highland birds; he had the dotterel, ring ouzel, redshank, oystercatcher, curlew, wheatear, dunlin, lapwing and golden plover. Although golden plovers were on the quarry list, he did not shoot them. Two other birds that give him great pleasure were the merlin and peregrine. Right on cue, at the end of my visit to Swaledale, a young peregrine flew in front of us and perched on a post.

It is on the question of birds of prey, particularly the hen harrier, that there is some conflict between grouse shooting and conservation. Most moors can cope with limited losses from peregrines, buzzards and eagles – or even the occasional sortie from woodland by a sparrowhawk or goshawk. But a hen harrier, with the taste and knack for taking grouse, can systematically search the moors and cause genuine problems. The hen harrier is a protected bird, and a very attractive one; rightly it cannot be killed or disturbed without obtaining a license from DEFRA or the Scottish Executive. In the case of 'economic damage' in England – English Nature advises DEFRA and their advice is always accepted. English Nature states that it is most unlikely ever to agree to a bird of prey being killed; consequently the farcical situation exists whereby a system has been set up to control the removal of birds of prey, in cases of undue damage, but no licence will ever be granted. Instead of real life, it sounds like something from the Mad Hatter's Tea Party. Yet, if a grouse moor fails to pay, because of hen harriers, and is given over to sheep or forestry, then not only is the hen harrier lost, but also the grouse, golden plover, merlin and all the rest.

The human reaction to death is interesting. If a rat kills a family of ducklings, the response is unanimous – kill the rat. If a film is shown of hyenas killing a wildebeest, the viewer is left in a state of horror and anger. But if a peregrine stoops from a great height; a hen harrier quarters summer moorland, or a kestrel falls like a stone – all we see is beauty – not the reality of death.

The reality of a bird of prey is death; it is a predator, it hunts and it kills. In my view if it is right (and it is) to control predators to preserve terns and wildfowl then there is also a case for controlling the occasional hen harrier on a grouse moor, where grouse numbers are low; especially if the alternative is forestry. Sometimes an identified rogue bird causes the trouble; it has simply become too good at killing grouse. This is rather like a 'rogue' fox that gets the taste for hens. Even the RSPB has had problems with rogue birds of prey in the past. Kestrels have developed a liking for avocet and little tern chicks in East Anglia. I believe that in all the instances those birds concerned should have been removed; they were not. Eventually the problem ceased, presumably when the birds died or moved. On an English Nature reserve there was another crisis involving a short-eared owl and sandwich terns; then to everybody's relief, not least to the sandwich terns themselves, the owl died for no apparent reason. I wonder how that happened? Luckily the incident did not happen on a gamekeeper's beat or there would have been serious investigations.

The RSPB does not agree with killing birds of prey: a spokesman said: 'I do not believe the RSPB could ever support the killing of a rare bird of prey. I can never see the day when the RSPB would support killing a bird of prey for grouse shooting. If we could resolve the conflict between harriers, gamekeepers and owners, it would be fabulous.'

The RSPB could resolve the conflict, almost overnight, but it will not, so it seems, because of a mixture of political correctness and a lack of honesty. One of the RSPB's common boasts is that its policies are driven by scientific research. Scientific research has clearly shown the damage caused to the red grouse of heather moorland by hen harriers. Langholm Moor in south-west Scotland was the area studied and the grouse bag fell from 1473 in 1992, to nil in 1998 as hen harrier numbers exploded. In 1999 the estate re-deployed its gamekeepers elsewhere. The tragedy of the 'Langholm Report' was that it only studied the affect of the hen harrier on red grouse – if it had added the golden plover, dunlin, lapwing, curlew and meadow pipit to the list it would have shown the true cost of hen harriers to conservation – a rare bird helping to create endangered birds. Just from its old English name 'hen harrier' it is clear that it has created problems for years, long before grouse shooting was ever thought of and when the loss of a hen or its chicks to an ordinary family represented a real financial and food loss. Ironically the most sensible solution to the hen harrier problem was put to me by an RSPB warden who despaired at the political correctness and practical conservation indifference shown by some of those desk-top conservationists at the RSPB's Bedfordshire headquarters. I met him when as presenter of BBC's then popular (popular here, in Europe and the USA) *One Man and His Dog* programme, we were recording a series on a hill farm in the North of England. The warden believed in a simple quota system, – allow one pair of harriers to breed on each estate and there would be enough grouse to satisfy the shooters: there would be golden plovers and curlew for the true conservationists and a number of hen harriers would be allowed to flourish for the politically correct conservationists. Sadly this sensible and practical solution was rejected by both the political and conservation hierarchies who seemed to put political and financial expediency before the welfare of curlews, lapwings, golden plovers, and yes, red grouse.

Unfortunately not only has this sensible solution been ignored but *One Man and His Dog* as a series has been lost too and I was sacked from the programme. I believe that my sacking came from my support for country sports, my views on Europe and for criticising the BBC's wish to 'dumb down' *One Man and His Dog*.

Another threat to red grouse and its habitat has, in the past been caused by extensive forestry in both England, Scotland and Wales – if only British foresters would go on an extended visit to Sweden to see forests managed in a way that can actually benefit many

The plough at work amongst the heather – with a new plantation in the background.

forms of wildlife. Regrettably the British forester seems to look on conifer monoculture in the same way as the wheat and barley barons regard their prairies.

In 1924 just 2.92 million acres were planted with trees; by 1985 the area had risen to 5.45 million. Since then around 1000 square miles have been planted – most with the alien sitka spruce and lodgepole pine. Britain needs timber of course; over 90% of our wood and wood products are imported. On private land most of this planting was achieved with tax incentives and grants, in much the same way as fifty years ago the same arguments and grants were being offered to cut down trees and rip out hedges to grow more cereals. Now, the haste and folly of these policies are widely accepted. Fortunately and belatedly it has been learnt that the policies allowing forestry in some of our most beautiful and unsuitable areas were misplaced.

When I first wrote *The Fox and the Orchid* forestry was being planted at an astonishing rate across spectacular areas of moorland. But conifer planting does not simply damage the land on which is takes place it can actually harm moorland close to it, affecting all the breeding birds, from grouse to golden plover and dunlin. The new dense forests harbour foxes and crows, causing a great increase in predation. So, as a direct result of afforestation, predation levels reach such a scale, that they can make adjoining grouse moors uneconomic. The increasing fox population is also thought to be an important factor in the alarming decline in the blackcock and capercaillie. A special survey led

Large scale afforestation. At least the steep area in the middle distance will eventually make an attractive break.

by David Stroud and Dr Tim Reed, has found that moorland birds avoid nesting within 900 yards of a forest edge, because of the threat of predators, which means that their breeding area is reduced still further. In addition forestry pushes out golden eagles, buzzards and ravens; it concentrates deer, causing over-grazing, and it reduces the hunting area of predatory birds, increasing pressure from peregrines and harriers on other birds.

There is a case for some forestry in the Highlands, as they once had extensive areas of forest. But they were natural forests, with glades and open spaces, allowing a variety of wildlife. Forests planted to Forestry Commission specifications are dark and dense and the only birds that seem to like them are coal tits, goldcrests and chaffinches. At the moment millions of people visit Scotland to experience its beauty, and the red grouse and red deer keep hundreds of thousands of acres of moorland free from sitka spruce. Few people would visit simply to see a giant Forestry Commission coal tit reserve.

The greatest excesses of planting were even carried out in the 'flow country', an area where until recently crofting, sporting estates and wildlife lived in harmony. Now, thanks to advances in ploughing and tree propagation, thousands of acres of land have been planted which are in fact totally unsuitable for mono-culture soft wood cultivation.

The 'flow country' gets its name from an Icelandic word for marshy moor and it is used mainly by English 'ecologists' who do not live in the area. It is a wonderful place of rolling heather clad hills, lochens, and blanket bog. Those who study bogs and 'mire' claim that only bogs in Ireland, Siberia and Tierra del Fuego match those of the Scottish flows. Several years ago I had a day watching falconry there, in Sutherland. A falconer flew peregrines over pointers after grouse. It was an unforgettable experience, complete with a distant view of two back throated divers.

In 1987 planting was taking place at the rate of 35 acres a day and 37,000 acres had already been planted. Because of the predators harboured in the new trees, bird populations had been depressed by at least a further 84,000 acres. The breeding territories of 640 pairs of dunlin, 1230 golden plovers and 260 greenshank had been lost in addition to numerous grouse. The future was bleak; forestry interests owned a total of 230,000 acres in the flow country; 20% of the total.

It was rather like the Tanzanian government ploughing up the Serengeti to grow barley but at least the Tanzanians could say that they were a Third World country – pleading a mixture of ignorance and economic necessity. As a First World country we were proceeding quite knowingly, without any pressing financial need; in fact, with poor quality timber as the end result, the whole exercise was madness – timber tycoonery encouraged by bureaucratic buffoonary. The legalised vandalism continued until 173,000 acres were planted. Now with the grants gone, sanity is returning led by the RSPB. From its 36,000 acre Forsinard Reserve the RSPB is buying adjoining blocks of forestry to remove the trees and return the area to a mixture of heather and bog.

The red grouse in its natural habitat. It is the grouse moor that is stemming the tide.

The flow country had a low grouse population and so sporting interests could not save thousands of acres from being planted with conifers. That I suppose shows the importance of the red grouse. It is not just a bird to be seen on bottles of a favourite whisky – it is a bird that if allowed to thrive is helping to safeguard our wonderful heather moorland heritage.

Chapter 8

Still Pool and Clean River

Although I no longer fish, I still look back to my angling days with much pleasure. It is not so much the fish that pleased me, but the surroundings in which I went fishing. The brook was my main fishery – with occasional visits to the River Cam. My memories are of hot afternoons drowsing on the brook bank, my float motionless among the lily-pads; dragonflies and damsel flies droning by; bees working the buttercups and daisies at my side, and sedge and reed warblers busily feeding their young in the standing forests of vegetation at the water's edge. Dusk at the dace pool was also a good time, with clouds of gnats attracting bats to hunt over the water, moorhens calling in the reeds and water voles munching noisily at succulent shoots. The attraction was the same as that which has drawn countless generations of fishermen to the banks of wide rivers and vast lakes, as well as to small streams and farmyard ponds. Izaak Newton caught the mood perfectly in his 'Angler's Wish':

> I in these flowry meads would be;
> These crystal streams should solace me;
> To whose harmonious bubbling noise
> I with my angle would rejoice:
> Sit here, and see the turtle-dove
> Court his chaste mate to acts of love;
>
> Or, on that bank, feel the west wind
> Breathe health and plenty: please my mind,
> To see sweet dewdrops kiss these flowers,
> And then wash'd off by April showers...

The fishing was good, with large roach, dace, perch, pike and eels – as well as great shoals of sticklebacks and minnows. There were huge, sinister-looking pike in the deeper pools, and I even had a hooked roach taken as I was winding it in. Because of their evil intent, we sometimes held 'pike shoots', with air rifles, to try and reduce their numbers – with little effect. The pike is a highly developed yet primitive killer, exuding malevolence; a distinction that it shares with few other creatures. H. E. Bates noted its menace, observing: 'The impression of streamlining, power and demoniac relentlessness given by his mouth, whether closed or open, is immense. A pike, indeed, arouses curious feelings of emnity for which I personally know of no parallel among English wild creatures.' I would have disagreed with H. E. Bates, for my old loathing for pike is still matched by my present dislike of rats.

Gradually fishing lost its appeal for me: I never liked forcing a barbed hook into a live worm. Finally I put myself off – by catching a number of magnificent perch in a deep hole on a bend in the brook's bed. They are among the most striking of our native fish, and each one had swallowed the hook deep into its throat. I felt guilty and have not fished since. If I had wanted to kill and eat the perch, it would have been different, but I could not justify pulling them out of the water 'for fun', simply to put back.

Others have felt differently, however, and the pleasure derived from fishing has been enjoyed by many generations of both countrymen and town dwellers. The most famous fisherman of the past was undoubtedly Isaac Walton and his book *The Compleat Angler*, published in 1653, is still in print and makes fascinating reading. In it, he quotes a poem by John Chalkhill, called 'The Fisher's Life', which sums up fishing's great appeal. It also helps to explain why Isaac Walton lived to a ripe and contented old age:

> Oh, the fisher's gentle life
> Happiest is of any;
> 'Tis full of calmness, void of strife,
> And beloved of many:
> Other joys
> Are but toys,
> Only this
> Harmless is,
> For our skill
> Breeds no ill,
> But content and pleasure.

Scenes similar to those created by Isaac Walton in words, were carved as woodcuts by Thomas Bewick, who caught perfectly the pastoral and reflective moods of fishing in his work. More recently H. E. Bates (1905–1974) identified another of fishing's great benefits: 'For there is no doubt that fishing cultivates patience, rather in the same way as walking stimulates thought.'

The junction of the old Test at Leckford – by Will Garfit. Tranquility is one of the special joys of a good river.

Just as he liked shooting (and foxhunting), Charles Darwin was a great fisherman. Indeed at one time his liking for country pursuits was so great that his father told him: 'You care for nothing but shooting, dogs and rat-catching, and you will be a disgrace to yourself and all your family.' Interestingly, like me, he also disliked the use of live worms. In his autobiography he wrote: 'I had a strong taste for angling and would sit for any number of hours on the bank of a river or pond watching the float; when at Maer (his uncle, Josiah Wedgwood II's home in Staffordshire) I was told that I could kill the worms with salt and water, and from that day I never spitted a living worm, though at the expense, probably, of some less success.'

Inevitably, in the past, not everybody liked fishing or approved of it. There is a saying which goes as follows: 'Fly fishing may be a very pleasant amusement; but angling or float fishing I can only compare to a stick and a string, with a worm at one end and a fool at the other.' (This is often, but incorrectly, attributed to Dr. Johnson.)

The fisher's gentle life.

In his 'To a Fish of the Brook', John Wolcot (1738–1819) was even more blunt:

Why flyest thou away with fear;
Trust me, there's naught of danger near,
I have no wicked hook
All coverd with a snaring bait,
Alas, to tempt thee to thy fate,
And drag thee from the brook...

...Enjoy thy stream, O harmless fish;
And when an angler for his dish,
Through gluttony's vile sin
Attempts, a wretch, to pull thee out,
God give thee strength, O gentle trout,
To pull the rascal in!

Today, rod fishing is Britain's most popular country sport (the pedantic divide 'fishing' into 'angling' – coarse fishing, – and 'fishing' – game fishing for salmon and trout. I use whatever word I think is suitable or most easily understood.)

There are about four million fishermen, and most of them are men (as well as boys), with enthusiasts still coming from both town and country. In the slow flowing lowland rivers coarse fish are the attraction; in clear chalk streams there are trout, and in the fast flowing rivers and streams of highland and moorland there are salmon, trout and sea trout. As if this was not enough, then sea fishing has a number of hardy followers and is a sport that is steadily increasing in popularity.

Once again, it is my view that the morality of fishing is something that should be left to the individual, as long as there is no deliberate cruelty involved, and as long as the sport does no damage to the countryside and waterways where it takes place, or to other forms of wildlife. For me, fishing for sport, as opposed to fishing for food, is the least acceptable country sport; but that is an entirely personal opinion, and I can see the appeal that fishing holds for others. Because of its place in the countryside and the fact that the decision to fish, or not to fish, should be a matter of conscience, I would find any legislation to ban fishing quite abhorrent, and totally counter to Britain's long tradition of individual freedom.

Otters – other traditional fishermen of the river bank.

A few anti-blood sports organisations are opposed to fishing, but in the main they claim to have no plans (at present) to press for its abolition. In the same way, most of those politicians who speak out vociferously against hunting with hounds, or even shooting, claim to find fishing totally acceptable. This suggests only one thing: as a similar case could be made against coarse fishing as against foxhunting or pheasant shooting, then other motives come into the anti-blood sport movement, and cruelty has little real place in the arguments.

The reasons are easy to see. Much of the opposition to blood sports has its roots in the Labour Party. Yet a sizable proportion of coarse fishermen are 'working class', from traditional Labour strongholds. Consequently parliamentary moves against certain country sports, but excluding fishing, must be seen as pure nonsense – a mixture of political opportunism and personal hypocrisy. It is summed up well by a piece of anonymous verse:

My anger simply knows no bounds,
When wicked Tories ride to hounds.
I'd rather watch the live fish dangling
When kindly Socialists go angling.

But beyond the arguments covering the ethics of angling, there is no doubt that fishing has played, and is still playing a very important role in conserving the health and quality of many of our rivers and streams, and of the fish themselves. The simple and obvious fact is that you cannot have healthy fish populations unless their whole environment is healthy. The link between clean water and the life within it and around it was easy to see in our stretch of brook before the Anglian Water Authority had its way. It was obvious too how all the elements depended on each other – the water plants, fish, insects and their water living larvae, frogs, toads, birds, fresh water mussels, the willow trees on the brook banks and many more. Then the water authority ripped out the bed; all the vegetation was removed, some fish starved, the rest disappeared and the brook died. It became nothing more than a drainage channel.

The struggle to bring back life to the brook has been a long and hard one. It has meant putting pressure on the water authority to let the bank vegetation grow without being cut back severely each year, and when they wanted to clear the bed again of obstructions and silt, it was done carefully, leaving water plants to grow at the side of the channel. They even co-operated by digging out a small pool, to vary the depth of the bed, and they left a number of gravelly shallows that had formed again quite naturally.

As a result, bankside flowers such as fleabane, hemp agrimony, forget-me-nots and marsh woundwort, are again plentiful. Beneath the water, beds of water starwort trail their bright green leaves, and sedges and rushes are returning. The presence of herons

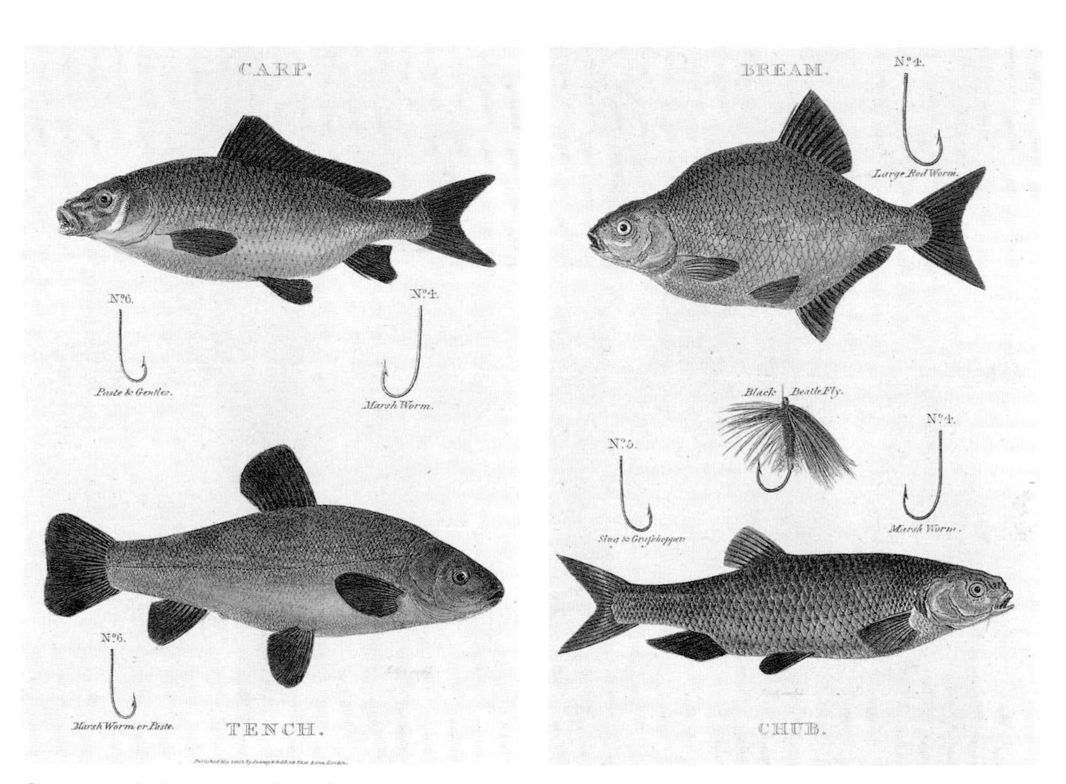

Carp, tench, bream and chub.

and kingfishers show that fish are back too – from pike at one end of the scale to sticklebacks and minnows at the other. The minnows can be seen at their best under a road bridge, sometimes in hundreds. In breeding condition they are spectacularly beautiful, as anybody with a net and a jam jar can confirm for themselves. Yet they are largely ignored, and assumed to be dull, brown and ordinary – 'tiddlers'. They have even been ignored by wildlife photographers and artists, and I have yet to see a picture that does justice to their miniature beauty.

They were not ignored in the past: In 1394, 7 gallons of them were served up at a banquet given by Richard II's chancellor, William Wykeham, Bishop of Winchester. Izaak Walton also liked them for their taste, as well as for their beauty, for after writing: 'The minnow hath, when he is in perfect season . . . a kind of dappled or waved colour, like to a panther, on his sides, inclining to a greenish and sky-colour, his belly being milk-white, and his back almost black or blackish,' he then added: 'In the spring they (fishermen and 'women that love recreation') make of them excellent minnow-tansies . . . fried with yolks of eggs, the flowers of cowslips, and of primroses, and a little tansy;

LEFT: *The presence of kingfishers indicates a healthy river or stream.*

thus used they make a dainty dish of meat.' Today they are welcome in any stream or river as they are a sign of clean, well oxygenated water.

Other fish to have appeared in the rehabilitated brook are shoals of large chub. When my nephew decided to try and catch one I must confess to being tempted to join him with rod and line, to see how large they actually were. If ever I succumb to the temptation it would have to be with a barbless hook. He caught one of over two pounds in weight. They can get up to ten pounds in British waters and even heavier on the continent. Since then, the water authority has 'electric fished' a stretch of the brook and confirmed the return of good sized pike, perch, dace, roach and chub. The eel now seems to be one species that has become scarce. Ironically, at one time it was thought that eels formed a significant part of the otter's diet, but from the number of spraints to be seen beneath the bridges where the brook flows under roads the lack of eels does not seem to have affected the return of the otter. Examination of the spraints by those who know suggest that the otter has switched its main diet to pike and chub. In the case of the pike most of the large fish appear to have become otter lunches – displaying either a remarkable turn of speed for the otter, or short staying power for the pike.

The status of mink along the brook is shrouded in mystery. In the eighties and much of the nineties mink caused wildlife mayhem – clearing the brook of moorhens and one was even seen reversing out of a kingfisher's nest with a young kingfisher in its mouth. Occasional mink are still seen, but some mooorhens have returned, suggesting that either mink numbers have declined or moorhens have become more adept at

Mink – the most pernicious of pests and sometimes released by animal activists.

avoiding these ferocious and efficient little predators. Some people believe that the presence of otters somehow suppresses the mink population – but with the otter depending almost entirely on fish and the mink surviving on one third fish, one third bird and one third mammal – especially rabbits – there seems to be little real competition for food. In addition there is a dramatic difference in size with a good male mink weighing up to four pounds and the females growing to just half that size. A male otter however, will weigh in at 22 pounds, meaning that with such a difference in size there will be little direct competition for breeding sites. If the otter and mink actually clashed physically, there would be only one winner and there would be plenty of signs of such a confrontation. No signs exist in Britain and in fact mink have been seen following otters in the hope of picking up left-overs from a successful fish catch. Although the threat from mink along the brook seems to have died down – other threats remain.

Sadly, the brook and its wildlife remain threatened by more than mink, for two days before the Christmas of 1985 there were dead and dying fish in the brook, with no signs of physical damage. The Environment Agency was certain that pollution had been caused by a farmer upstream, either spraying fields close to the brook, in windy conditions, or through washing out his sprayer and allowing the poison to seep into the brook. Both could have been true as some farmers had been spraying, despite there being a high wind. Officials could go even further and identify the man responsible – off the record. He was not prosecuted for tests had to be undertaken at a laboratory to establish proof, yet remarkably, the scientists and officials had already stopped work in anticipation of Christmas.

It is a surprising and depressing fact that the largest number of pollution incidents each year are now coming from farms – sprays, and seepage from slurry tanks and silage pits. It is in the fight against pollution that fishermen have played a vital role. Their motives have been quite easy to understand; if a waterway is polluted it has no fish and so they cannot enjoy their sport. But a river or stream that is too dirty to support fish is also incapable of supporting other kinds of wildlife, from water-lilies and 'pokers', to dragonflies and kingfishers. So when anglers fight for their sport, their actions can benefit far more than fish.

Fishermen first joined together to improve their fisheries as long ago as the 1880s. Then, coarse fishermen from London and Sheffield urged one of the Sheffield Members of Parliament to press for a close season for coarse fish, to protect them during the breeding season. Protection was given and various other Acts of Parliament such as the 1923 Salmon and Freshwater Fisheries Act, resulted from the work done, and the pressure applied, by the anglers.

Possibly the most important step taken was in 1948 when a lawyer, Mr John Eastwood, formed the Anglers' Co-operative Association (ACA) to use the Common Law to ensure that a downstream riparian owner received unpolluted water. Today the ACA has a membership of 8000, together with 1000 affiliated angling clubs. As a result it keeps a constant eye on water quality throughout Britain and it has the resources to take legal action against those responsible for incidents of pollution.

Its most famous case took place in 1951 when the ACA, on behalf of the Pride of Derby Angling Club and the Derby Angling Association, took action against British Celanese, Derby Corporation, and the East Midlands Tar Distillers, who were dumping millions of gallons of untreated sewage and various tar products into the river Derwent, which in turn was polluting the River Trent. After thirteen days in the High Court and six days in the Court of Appeal, the case was won, with costs and damages awarded against the polluters.

In 1953 successful action was taken against the huge steel firm of John Summers Ltd on the River Dee, and in the following year the Consett Iron Company was prosecuted. A successful ruling was even obtained against Sevenoaks Council, with the judge threatening to send the councillors to prison if the pollution by sewage did not stop. Shortly afterwards the council built a new sewage works. Since then the ACA has handled many hundreds of cases resulting in hundreds of thousands of pounds being awarded in costs and damages, to allow fishermen and riparian owners to restore their fisheries. Despite this there is evidence that in some areas water quality and consequently water life are again declining and so the work of the ACA is far from finished.

It does seem remarkable that the ACA should be required to take companies, individuals and even local authorities to court under the Common Law in the first place. It is surely time that environmental standards were both high, and maintained, under the Criminal Law, so that all polluters were automatically prosecuted by the police or the appropriate government agencies. It is almost beyond comprehension that a healthy countryside should have to depend on the vigiliance of private individuals.

The passing of the Control of Pollution Act of 1974, with several additions since, has improved the position enormously; credit for this must be given to the angling lobby. But there are still loopholes and the ACA's work continues. Without it there is little doubt that many of Britain's waterways, including several canals that have become good fisheries, would have been virtually dead – killed by various industries and local authorities using them as cheap, open drains.

The influence and importance of fishing in the creation of rich wildlife habitat goes far beyond coarse fisheries and flowing water, for all over the country there are countless ponds, pools and lakes that fishermen have saved and restored. If an old pond or gravel pit is left untouched it slowly dies. It is a natural process; weeds grow, willows and alders block out the light, the water becomes black and stagnant. But if a pool is fished, it is 'managed' – light is allowed in, trees and plants are controlled and conditions for a whole range of wildlife are created.

In areas of intensive agriculture these places take on great significance, forming small havens and refuges in an overall hostile environment. Two such places close to me in Cambridgeshire illustrate the point very well. One small lake actually forms part of the water system of Wimpole Hall, the large country house once owned by the daughter of Rudyard Kipling, and now in the hands of the National Trust. It was a feeder lake and fell into disrepair; for forty years it remained neglected; its sluice became useless, most of the water drained away, and it became a boggy tangle of reeds and willows. It was then bought by a local farmer, who gradually transformed it; the bed was cleaned out and the area of water was enlarged; the willows were removed, an island was created and the sluice was repaired. The dense woodland surrounding the lake was also tackled with some of the undergrowth and scrub removed to let in the light.

Today it is a small wildlife paradise. Beneath the surface of the water it is alive with various invertibrates, water fleas, mites, leeches, diving beetles, various water snails – including the attractive ram's horn and pond snails – freshwater shrimps, water boatmen and the larvae of the caddis fly, damsel fly and dragonfly. There are underwater plants too, with stonewort, water milfoil, Canadian pondweed, and many more. The edges provide a rich area for plants including various sedges, rushes and reeds, as well as attractive flowering plants – amphibious bistort, the white water-lily, water crowfoot and water plantain. On the banks there are wild strawberries, violets, cowslips and creeping jenny. The surrounding wood is of old hawthorn, ash, field maple, oak, willow

Sea clubrush at Will Garfit's gravel pits – the water-edge plantlife is essential for fishing and wildfowl interests.

The small wildlife paradise, near Wimpole Hall.

and a wayfaring tree. Some of the oldest trees have holes, allowing woodland birds to nest successfully, including all three woodpeckers — the green, lesser-spotted and great spotted. Footprints (more accurately called 'slot marks') also indicate other visitors to the lake, for both fallow and muntjac deer find its seclusion attractive.

A visit in late May can be perfect, with swallows and house martins feeding over the water; sedge and reed warblers in the rushes, and from the trees come the melodic songs of willow and garden warblers. Blue damsel flies skim the water lilies and a little grebe sits on her nest. Rings of ripples show fish rising, for not only has the lake a healthy population of indigenous rudd, but it has also been stocked with rainbow trout for fly fishing.

As a result, an overgrown and stagnant reed-bed has been turned into a lake of beauty and richness. It provides a haven for nature, pleasure for fishermen and an income for the owner.

Not far away old gravel pits have undergone the same change. Once they were owned by local builders – but then they were sold. Some of the pits were of open water – others were totally overgrown. For the shrewd businessman they could have been purchased and used as a tip for land fill – a very remunerative option. Instead they were bought by Will Garfit, the artist, some of whose excellent paintings and sketches illustrate this book.

Through his care and understanding of wildlife, and the landscape it needs, the old gravel pits have become an attractive area of abundant life. He has created new pools; he has restored many of the old ones and he has introduced marsh and water plants, both to improve the quality of the water and to provide seeds on which wildfowl can feed. He has planted woodland, opened up some of the old thickets, and cut clearings. He has even left some old, dark tangles of willow to create a complete cross-section of habitats to encourage a maximum range of plant and animal life.

He is being successful, for on a July day, although patches of common spotted orchids were just over, there were masses of willow herb, fleabane, century, silverweed, stitchwort, selfheal and dewberry blossom. Wrens, willow warblers and whitethroats sang and house martins fed, low over the water. Broods of greylag and Canada geese cruised contentedly, and moorhens and coots were busy in the reeds. It was good to see moorhens and the pits attract kingfishers, too. Butterflies were also active in the sun – small tortoiseshells, gatekeepers, meadow browns and small skippers. Michael Chinery, the naturalist and author of several good wildlife books, visits the pits from time to time. So far he has recorded seventeen species of butterfly, which is good for the area. Like me, he does not hunt, shoot or fish, but he says: 'Without sporting reserves such as Will Garfit's, Cambridgeshire would have a lot less habitat for wildlife – and every little reservoir is vitally important . . . early results (moth trapping) indicate a pretty rich moth fauna – not unexpected in view of the diversity of habitats within the reserve, but encouraging when one realises its isolation from other suitable habitats.' The old gravel workings cover 70 acres in all and Will has turned them into a successful nature reserve, shoot (he rears a few pheasants) and fishery, with carp, rainbow and brown trout.

The brown trout is Britain's indigenous trout and the best fishing for it is found along fast flowing, clear streams. Those who fish prefer to fly fish on chalk streams. Water flowing from chalk is known for its clarity and purity and in years gone by chalk streams were noted for their water meadows and watercress beds. The wildlife is important, too; grey and yellow wagtails, herons, moorhens and kingfishers are typical birds, while the 'beautiful demoiselle', one of our most striking damsel flies, is an insect of clean, fast flowing streams. The water life includes other fish that only thrive in clean, pollution-free water; the small stone loach, the strange and primitive brook lamprey and the grayling, one of the most sensitive fishes to the presence of pollution and beaten in beauty only by the salmon. The crayfish is another that favours healthy chalk streams – it is unmistakable, looking like a miniature freshwater lobster. Again, trout fishermen have been important in preserving the richness of many of our streams and rivers; guarding against both pollution and loss of water level caused by the authorities extracting too much water for our thirsty and extravagant cities and towns.

My late and greatly missed friend Gordon Beningfield was another artist who had a long love affair with trout fishing. I first met him when he was involved in a television programme, and he came to interview me about my tame fox which I had at the time – we remained friends until his death in 1998. I had many memorable experiences with Gordon; I was with him, doing an interview for *Woman's Hour*, on BBC Radio, when I first saw the parrot along our brook meadows. Then I went as moral support with him to the RSPB's headquarters at Sandy in Bedfordshire, where he berated the Director for concentrating resources on special reserves and not farmland and the general countryside. I went with him to book signings in Cambridge too, as he claimed that he could not spell – unfortunately for him I can't spell either. On another occasion he suddenly sprinted off, as I was talking to him outside his cottage; he had seen a youth surreptitiously poaching trout with a worm and handline, and set off in pursuit. He loved to laugh; on seeing a linnet he would always say: 'Look there's a bush with a linnet init. It is a linnet in'it?'

When I first knew him, he was an enthusiastic pheasant shooter and trout fisherman. But he gradually changed; his interest in the countryside for its own sake increased so much that he had doubts as to whether he would ever shoot again: 'I think differently,' he said. 'Now I get more enjoyment out of looking.' It was a change experienced by Charles Darwin, too; he wrote: 'I discovered, though unconsciously and insensibly, that the pleasure of observing and reasoning was a much higher one than that of skill and sport.'

But Gordon still thought country sports were important: 'Whenever the sporting interest enhances the landscape for wildlife, it is to be encouraged. It should be obvious to everyone that there is more wildlife on sporting estates than on straightforward commercial farms. In saying this, though, I disapprove of heavy predator control – predators must be tolerated. Fortunately there are enlightened gamekeepers these days – although unfortunately some of the old backwoodsmen are still around, too.'

Although his shooting virtually finished, in the summer of 1986 he invited me over to see a stretch of the River Gade, near his home in Hertfordshire, on which he still had a rod. The river flows from the nearby Chiltern Hills and is crystal clear. Water meadows with grazing cattle accompany it; as we walked and talked some cattle broke through a fence into fresh grass. It was good to see it happening on a Wednesday – our cattle always seem to get out on a Sunday.

There were clear pools, with great weed streamers of many shades of green; and tinkling shallows with flowering water crowfoot and stands of yellow iris. Coots, moorhens, mallards and minnows were all plentiful and butterflies flopped lazily. It was good to find grassland that still had grasshoppers below, and skylarks singing above. Despite the closeness of a busy main road, it had the feel of an England that in many

Restocking trout in the River Wey.

areas has been brushed aside by time and 'progress'. Old watercress beds were overgrown and humming with life and willow warblers were singing their tumbling song of summer. Occasionally a trout would rise. But what started out as a discourse on fly-fishing gradually turned into a ramble – an old fashioned nature walk – smelling flowers and watching butterflies – all because of a clean, healthy river. During the 1986 season the other rods enjoyed their sport, but Gordon managed to go the whole time without catching a fish: 'I did go out with my rod several times,' he claimed, 'but I always seemed to get interested in something else.' Interestingly, Gordon's trout fishing friends later agreed that the river landscape and all it's wildlife was to be a priority in the aims of the syndicate.

But although the contribution made by angling to clean water is obvious and beneficial, it has to be said that fishing also causes a number of problems. Many keen and responsible anglers are aware of the difficulties and try to lessen them, but there are others who simply could not care less. The start of the fishing season (June 16th) coincides with the main month for breeding birds. Yet some fishermen will cut down reeds and rushes, to get to the best fishing sites, disturbing and sometimes damaging nests. Even if no actual damage is done, disturbance takes place on a very large scale. As a result it seems only obvious that the close season for coarse fishing should take into account all the creatures of the water-side, not just fish, fishing ought not to be resumed until July 16th. This would enable warblers, grebes, waterfowl and kingfishers to rear

The brilliant camouflage of the brown trout among starwort in the Test.

their young in comparative peace. Where otters are present it even seems sensible to limit fishing to just one bank, to reduce the amount of disturbance. Such changes would be resisted by many fishermen, but they are sensible and long overdue. Coarse fishing is far more disturbing than fly fishing; the coarse fisherman is static and can keep a bird from its nest for hours at a time and if a fishing competition is taking place with fishermen posted every few yards whole stretches of riverside experience day long disturbance. The fly fisherman on the other hand is often solitary and moves along a stretch of water looking for his quarry.

Litter is another problem caused by the irresponsible. It is particularly noticeable after fishing competitions, when urban anglers will often leave their beer cans and assorted debris, apparently unconcerned about the dangers they can create for water birds and farm livestock.

Discarded tackle is a huge part of the problem. Between June 1978 and March 1979, members of the Young Ornithologists' Club (the junior section of the Royal Society for the Protection of Birds) recovered over six miles of discarded fishing line and 3000 pieces of split lead shot. Among the 42 birds reported as being found dead or injured as a result of becoming entangled in fishing line were five Mute Swans. Thirty miles of river banks produced an average of 808 feet of line, 86 pieces of split lead shot and seven hooks every mile. Other equipment found included four lead weights, five floats, one disgorger, one swivel, one spinner and seven flies.

Their report acknowledged that observers inevitably chose water margins which were most used by anglers and also stated: 'Because of the difficulty of finding split lead shot in vegetated areas, the average of 86 pieces per mile is probably only a fraction of the total actually present.' Further surveys were carried out in 1982, 1989 and 1994. Unfortunately the distances and places covered vary enormously – as did the type of measurement – changing from miles to kilometres – but the findings do show that the amounts of line, hooks and weights lost or discarded have declined considerably – helped by the outlawing of poisonous lead weights (replaced mainly by tungsten) in 1987. In 1994 over a distance of 23 kilometres, four metres of line were found per kilometre and a total of 58 lead weights, 58 tungsten weights and 39 hooks. On this basis along 56,000 miles of river in Britain – there could be 560 miles of abandoned fishing lines and over 152,000 hooks left to endanger birds and animals of the river bank.

Of course this problem is not restricted to the immediate waterside, or to Britain. While visiting northern Sweden in 2003 with Britain's only reindeer herders – Alan and Tilly Smith – we found a long length of line 200 yards away from a river. It was in a beaver run – as if a beaver had become entangled and carried the line well away from the river bank. With otters, mink, ducks, herons and swans discarded line and hooks can be taken long distances from the water's edge causing untold suffering and long lingering deaths.

Each year animal refuges and hospitals have numerous birds, particularly swans, with

Trolling for pike – by James Pollard.

horrendous injuries caused by hooks, spinners and line. In the first edition of *The Fox and the Orchid* I described the heart rending scenes at a Swan Rescue Service in Norfolk run by Ken Baker. He eventually had to give up as the stress and trauma were just too much for him If fishermen only took note of the Nylon and Litter Code of the National Federation of Anglers' the problems caused by lines and hooks could be reduced still further. An additional reduction could be achieved if young children were given more help and supervision when they started fishing.

I had many happy hours fishing as a boy, but once again I think that today's predominantly urban outlook brings a certain disregard from city youngsters, both for the fish themselves and for the river bank in general. Consequently I believe that the time has come when boys under fourteen should not be allowed to fish without supervision from a responsible adult, and if they have no adults readily available, they should join a fishing club for the necessary help and tuition.

The National Anglers' Council already runs a good Proficiency Award Scheme for young people over the age of eleven. It covers all aspects of angling, from the actual skills needed to catch a fish and the correct way to handle it once it has been caught, to conservation, the removal of litter and responsible behaviour along the river bank. Perhaps The Environment Agency should insist that all new fishermen should pass through the scheme before a full fishing licence is granted?

Chapter 9

Where Salmon Leap

Just as hedgerows are of great importance in lowland Britain as wildlife corridors, giving birds and animals cover, shelter, food and access over large areas, so many of our most attractive rivers can have a similar function. A clean, well kept river is good for wildlife, and its valley can form a natural refuge all the way from its source to the sea. If that river is clear, with a good flow of water over gravel and rocks, it will provide a highway for dippers, grey wagtails, kingfishers, and, in the North, Wales and Scotland, the common sandpiper. It provides a habitat suitable for the otter, too.

Over recent years I have visited many such rivers, including the Exe, Dart, Camel, Tamar, Test, Itchen, Severn, Hampshire Avon, Frome, Tweed, Spey, Dee and Findhorm. But as I have walked along their banks I have hoped for more than the birds and flowers of flowing water; I have looked in the pools for a glint of silver, or a leaping, streamlined form slicing through white-water and jumping at the foot of falls. For these rivers are famous for their salmon; allowing them to travel along fresh-water to breed, away from the sea.

Isaak Walton called the salmon 'the king of fresh-water fish', and without a doubt the salmon is the largest and most beautiful of all our fish. Even now I feel a surge of excitement every time I see one, as the story of the Atlantic salmon, whose name comes from the Latin *salar* (the leaper), is remarkable. Its cycle starts in the clear, remote headwaters of rivers and streams where the adults lay their eggs in the autumn – to hatch the following spring:

> In the gravel of the moorland stream the eggs were hatching, little fish breaking from confining skins to seek life, each one alone, save for the friend of all, the Spirit of the waters. And the star-stream of heaven flowed Westward, to far beyond the ocean where salmon, moving from deep water to the shallows of the islands, leapt – eager for immortality.

So Henry Williamson finished his beautiful book *Salar the Salmon* – capturing the entire spirit of the salmon's romantic life cycle in a single paragraph.

Two years after hatching, the young salmon, 'smolts', swim down to the sea, where they make for their feeding grounds, rich in herrings, sand eels and sprats. For years it was not known where they fed, but then neaarly fifty years ago their feeding grounds were found off the coast of Greenland and around the Faroes.

The salmon stay at sea for two years or more before returning to their home rivers, although some 'grilse' return after only one year. What guides them back is unknown; perhaps the stars, moon, position of the sun, magnetic influences, or simply the rhythms and currents of the sea, although there is now some evidence that at the end of their journey they identify their home river by the smell of its freshwater.

Twenty-six years ago I saw salmon in the Findhorn estuary on the east coast of Scotland. It was a place of white sand and ice-blue water where the river flowed gently into the sea. There, as the fish ended their ocean journey, local men were using a sweep net and rowing boat (a net and coble), to make broad sweeps of the river after the fish. They had caught three, the largest being about 10 lb of gleaming, streamlined silver. But they were disappointed: 'Ten years ago we would have caught 20.'

Later, while walking by the river Spey I saw a salmon arch into the air, with sparkling droplets of water cascading from its glistening, plunging body. It was an exciting sight, for the fish had run the gauntlet of boats, seals, nets and rods, and would soon be at the river's source. A fisherman with his rod and line complained that he had seen few such fish on his fortnight's holiday. Unfortunately his experience is not unusual, for over several years the number of salmon returning to our rivers has been steadily declining. As another cynical fisherman informed me: 'It's the spring run which has declined the most alarmingly. By the time the fishing MPs get here in late summer the grilse are running – they get good fishing and think we are crying wolf.' Other fishermen in Scotland have the same story. One who has been helping clients on the Dee for many years told me: 'Now we are taking two fish in a week, where 40 were taken in the 'sixties.'

Lord Home, the former Conservative Prime Minister who died in 1995, agreed. He loved fishing, and had fished since he was a boy. His home river was the Tweed and there, too, the spring run had almost petered out. He believed that it was because the character of the river had changed, with drainage from hill farming and forestry up-stream, and also because foreign sea-fishermen had found the ocean feeding grounds of the spring salmon. Unlike many rivers, however, the Tweed's autumn run was getting better, reversing the general trend and confounding the prophets of doom. The autumn of 1986 beat anything he had ever experienced in a long and happy fishing life: 'Looking over the bridge at Norham, six miles above Berwick, I saw between 50,000 and 100,000 salmon, with no exaggeration whatsoever – stretching for a mile and a half. The surface of the Tweed was foaming with fish, splashing about like those old photographs of the herring nets, and there were some great, whacking brutes. There had been a drought and they had not been able to get up river. I had not seen anything like it before. It was incredible; I could hardly believe it. The autumn run for the last five years has been tremendous.

'I can remember the old autumns which were marvellous; but now it's even better. The fish are slightly smaller, but this year (1986) it's confounding that – they're even bigger.'

Lord Home's knowledge of fishing and the pleasure it gave him were quite evident. He was quite unlike the image protrayed of him in the political media. Although as a boy he started fishing with a hook, line and worm, he loved fly fishing best of all: 'If you are fishing a floating line, the excitement of the fish coming up and taking it is tremendous. Then the line sinking and the thrill of it going up your arm when you come in contact with a living animal — unseen under the water — it's tremendously exciting. In fishing there is nothing of deliberate cruelty. I am not morally worried about fishing. I think fish are one of the foods placed by the great Creator for human beings to eat.'

It seems to me, too, that salmon fishing with lure or fly has no ethical or conservation objections. There are fewer salmon anglers per length of water than coarse fishermen, leading to little disturbance, and many of the hooked fish are killed as soon as they are hauled from the water, to be eaten. As fishing with rod and line is an inefficient method of catching salmon, it also allows a potential breeding stock to move upstream to the spawning grounds. However, to increase the number of fish returning to their spawning ground some rod fisheries are now practising a 'catch and release' policy. In a trial carried out by the Atlantic Salmon Trust's Field and Research Biologist, John Webb, 80% of fish caught in the spring by anglers, handled carefully, radio-tagged and released, were detected in the spawning areas at the end of the year. It is estimated that 39% of all salmon caught by rod and line in Scotland in 2002 were released.

Sadly, most of the British rivers (and those of Europe and North America) are not like the Tweed, and the reality is clear; in the last thirty years the North Atlantic catch of wild salmon has declined by 80%. There are many reasons for the fall, which in total pose a considerable threat to the future well-being of the salmon. The real problem started in 1966, with an outbreak of a strange disease, ulcerative dermal necrosis (UDN). It is thought to be caused by a virus; the fish become ulcerated and infected with a fungus. There is no cure. The disease appears to run in cycles, rather like myxomatosis in rabbits; its arrival led to a rapid decrease in salmon numbers, and although it has declined, it is still present in many rivers.

The discovery of the salmon's feeding grounds around Greenland and the Faroes was also a disaster. It immediately led to new salmon fisheries being set up and the inevitable over-fishing; fortunately quotas are now set, in an attempt to preserve stocks. Many fishermen consider that the fishing of the feeding grounds only affects those salmon taking part in the 'spring run', as several rivers, in addition to the Tweed, still experience a good 'autumn run'. It is assumed that the feeding grounds of the late running fish are still unknown.

Those fish that survive the nets and hooks off Greenland and the Faroes have more problems when they reach British waters. Although drift-netting was banned in Scotland in 1964 because of the threat it posed to Scottish salmon, it still takes place off the north-east coast of England. Many thousands of fish are taken, 94% of which, it is estimated, are making for Scottish waters. The fishing is undertaken with nylon monofilament nets, which injure many uncaught salmon, causing them to fall victim to disease and predation. Both the RSPB and the World Wildlife Fund are concerned by this form of fishing, as it takes place along part of the coast rich in many forms of wildlife, at the height of the summer breeding season. Each year it is thought that the nets kill numerous guillemots, razorbills, puffins, gannets, shags, cormorants and diving ducks. At other times of the year red-throated and black-throated divers can be caught; it is also common for seals to be killed and injured in the nets.

Fortunately the North Atlantic Salmon Fund (UK), (NASF (UK)), has recently signed an agreement with some of the Netsmen of the North East Drift Net fishery. As

The salmon – the most beautiful of all fish – by Rodger McPhail.

a result 52 out of a total of 68 Netsmen have agreed to surrender their licences and stop drift net fishing for salmon and sea trout. The payment to the fishermen will be completed in May 2004. This represents 76% of the licensees who catch salmon and sea trout from Whitby to Holy Island and it has been estimated that 80% of the normal catch will be saved. This will enable fish to return to their native rivers in Yorkshire, Northumberland and along the East coast of Scotland

The annual catch of the fishery between 1993 and 2001 was 33,655 salmon and 33,907 sea trout; consequently an estimated 26,000 of both salmon and sea trout should be free to return to their home rivers. The total cost of the payout will be £3.34 million of which £1.25 million has been provided by the government. Rivers such as the Tyne, Tweed, and Tay should be among the first rivers to show an increase in the numbers of returning fish.

Sadly this still leaves a significant drift net fishery off the west coast of Ireland that takes large numbers of salmon and sea trout returning to rivers in Scotland, Wales, north-west, south-west and southern England and France.

Unfortunately much drift-netting for salmon takes place illegally, too. The salmon are brought into harbour with white fish placed on top for concealment. I was told by one conservationist on the west coast of Scotland that the practice was widespread. 'But what do I do?' he asked. 'The fishermen are my friends, foreign trawlers are taking their other fish, and if I report them, I lose my salmon too.'

At one time it was thought that salmon farming would come to the aid of the wild salmon. With a plentiful supply of cheap farmed salmon there would be no market for wild salmon. Sadly, the market for wild salmon continues, as it is considered to be a better 'product', and the environmental impact of farmed salmon on their wild cousins has been little short of catastrophic, particularly along the west coast of Scotland.

With the huge numbers of salmon, confined in sea rearing cages have come high concentrations of sea lice which attach themselves to the fish and can kill them. Migrating salmon and sea trout passing close to the farms are very vulnerable to attack, and some west coast populations have been virtually wiped out. Incredibly up to 100 lice can be found on a single six inch smolt (young salmon migrating from fresh water to the sea).

Where the infestations are treated chemically – some of the toxins, including organo-phosphates, can then create both environmental and biological problems. Indeed some food experts believe that as a result of the chemicals administered both externally, and internally – fed as treated pellets, those eating large quantities of farmed salmon put themselves at risk. Yet more pollution is caused by the wasted food and fish excrement that falls below the cages.

Incredibly it takes eight tons of industrially fished small fish – including sand eels – processed into food pellets, to create one ton of farmed salmon. This represents a huge loss of sea life, used inefficiently, simply to produce cheap smoked salmon. As I write (July 2003) seabirds are dying in their thousands around the Shetlands as sand eels have 'disappeared'. The 'experts' and the politicians simply cannot think where they have gone – oh what a surprise.

Escaped farmed fish can cause further problems. Local fish are adapted to their home rivers to aid survival. When interbreeding takes place the overall fitness of the young to suit their local conditions is affected and survival rates fall. In addition to all this high densities of farmed fish can lead to the spreading of infectious diseases and parasitic infestation such as Infectious Salmon anaemia and Gyrodactylus salaris.

Poaching is common on the rivers as well, with nets, gaffs and even cyanide. Sub-aqua lungs, harpoons and inflatable boats are also being used as the unacceptable, large-scale commercial poacher replaces the likeable local rogue, taking the odd fish to supplement his earnings, or beer money. Tagging salmon carcasses, giving each one an identity, could help combat poaching, as it has done successfully in parts of Canada. Sadly, and surprisingly, the politicians seem to be against this, although many conservationists and fishermen believe that the real opponents are the civil servants. As Rear Admiral John Mackenzie, vice-President and former Director of the Atlantic Salmon Trust told me: 'If the will was there I have not the slightest doubt that the salmon tagging scheme could be made to work. Unfortunately it does appear that in government departments in England and Scotland, there is not the will to make it work.' Others are not so polite: 'The truth is that the civil servants here are too bone idle to get it going; they argue why it won't work instead of seeing the reasons why it will.'

In an attempt to deter poachers under the Salmon Act 1986 the government has made it an offence to be in possession of a salmon not caught legally. Only licensed dealers are now entitled to buy and sell salmon. It seems another piece of legislation designed to provide good perks for the legal profession but few actual prosecutions. There are other predators too that cause problems; seals, cormorants, red-breasted mergansers and gooseanders, that could all be controlled sensibly to help the survival of the salmon. Unfortunately as with all predator/prey problems emotion and political correctness is currently being put before consvervation.

Pollution is yet another element in the reduction of salmon numbers. In south-west England and Wales there have been several instances of farm slurry causing the large-scale loss of both adult and young fish. In Scotland and Wales things are more sinister. There is evidence that an increase in acid rain is causing the gradual death of several Scottish rivers and lochs, and young salmon cannot survive in water that becomes too acid. The pollution is being caused by the discharges of coal and oil fired power-stations falling back to earth, dissolved in rain drops. Indeed some people think that acid rain could be one other factor in the decline of the grouse and all upland life.

The acidification of rivers is also being adversely affected by blanket afforestation. This is caused by nutrients from the soil being leached out and also by the deposit of pine needles. The acidity of water is measured as a 'pH' value; the lower the number, the greater the amount of acid in the water. With spruce being planted, it has been known for a pH value of 7 to fall to 3.2. Young salmon cannot survive in water with a pH value lower than 4.5. So, not only is widespread forestry reducing moorland – it is gradually killing some of our highland lochs, rivers and streams as well.

The drainage of newly forested areas or improved farmland can have two more serious affects on salmon rivers. Erosion is often caused, depositing silt on the gravel spawning beds ('the redds') making them unsuitable for the salmon to breed. The drains change the river flow; the old river depths were based on seepage, giving steady and reasonable water levels. Now, after rain, the water flushes away quickly, causing rapid and significant fluctuations between high water and low water. This gives the salmon less time to move up stream and washes away the ground on which they have spawned, causing local flooding. Large scale tree planting near important rivers and streams also causes a fall in the water table and a serious reduction in flow.

The River Locky with Ben Nevis behind – by Will Garfit. This view makes the salmon fisherman reach for his rod.

Fortunately not all is lost, as the fishing interests have been working hard to overcome the problems. Dr Derek Mills formerly of Edinburgh University has had a long term interest in salmon and has done much work on the management of forest streams to reduce the harm done to aquatic life, including salmon – young and old. He believes that it is important to try to ensure well oxygenated and silt-free water, despite the difficulties. To achieve this all streams should be regarded as 'reserves', and a distance of ten times the width of the stream, on each bank, should be given priority treatment. Pine trees must be kept well away; there should be plenty of light allowed in and natural vegetation encouraged, including small deciduous trees such as birch, willow, rowan and alder. The willow's root system is particularly good at helping to prevent bank erosion, and the alder leaf has been found to contain four times as much nitrogen as other deciduous species, and so is good for fighting acidity. At the same time the breeding grounds should be checked regularly and kept clear of silt and any debris washed down by badly constructed drainage ditches.

Water problems of another disturbing kind affect the sea itself. Changes seem to be occurring in sea temperature, salinity and the ocean currents themselves, which in turn affect the production of plankton and the whole food chain connected with it – thus affecting the salmon's chances of survival at sea. These changes are almost certainly connected with long term global warming – an issue threatening far more than simple salmon survival; it has huge implications for humanity as well – something our short-term and dim-witted politicians have yet to come to terms with.

Work is also taking place on the salmon itself. The Atlantic Salmon Trust has a scheme of radio tagging, to monitor the movement of fish entering the Dee, in order to help our understanding of the salmon. The Atlantic Salmon Conservation Trust on the other hand has undertaken an exciting scheme, which is still continuing, to help improve salmon numbers. It is buying up the rights for netting salmon on their coastal migratory routes – it is called interceptory netting by 'fixed engine' (stake or bag nets). It also wants to buy out sweep nets in the estuaries, in an effort to get salmon back to their rivers of origin. So far it has cleared the nets from the rivers Conon, Beauly, Ness, Nairn, Findhorn, Spey, Dee, Don and Tweed, at a cost of over £2 million and the process is continuing. All the purchases have been concluded on a willing-seller, willing-buyer

The newly forested Highlands do not help the spawning grounds of the salmon.

basis, with all the rivers' fishing owners, including the Balmoral estate on the Dee, contributing to the fund. The removal of over 100 fishing stations has meant that at least an extra 100,000 salmon are now returning to their rivers. Over time this means that many more mature fish will be returning to their spawning grounds and with the removal of the drift nets too – gradually the population of the Atlantic salmon, one of the most remarkable and beautiful fish in nature, should begin to improve. All this should be viewed in economic terms, as well as conservation terms, as salmon fishing by rod and line is very important to the economy of rural Scotland, and is worth far more in money, employment and services than the coastal fixed engine fisheries. One netted salmon is said to be worth £20 to the local economy. One large salmon caught by rod and line is said to be worth up to £5000. Not all Scotland's bag, stake and sweep nets are to close, as they do form an interesting and traditional part of Scottish life. Their reduction, however, means that the pressure on the salmon is lessened.

Without the efforts of the freshwater salmon interests it is highly likely that with fishing at sea, nets at the estuaries, pollution, predation, and the affects of afforestation, the salmon would now almost be extinct in British rivers.

In fact the salmon has almost become a status symbol for clean water. Salmon have successfully been re-introduced to the Thames, and a similar scheme is being tried with the Trent. Both were once good salmon rivers, but they died with the pollution of urbanisation and industrialisation. Although some salmon do now return to the Thames, and much is made of the clean-up of Britain's 'greatest river', to my rural nose and eyes there is still a long way to go. It is interesting to note that the largest tributary to the Thames is said to be a twin sewage outlet near Tilbury.

After all the despondency, there is a positive air about the Atlantic salmon. Jeremy Read, Director of the Atlantic Salmon Trust says: 'I think that there is still cautious optimism about the future of the wild salmon. Whatever factors may be outside our control, the survival of this wonderful fish is no longer taken for granted and there is increasing co-operation in tackling the problems that can be attacked. The various organisations who care about the wild salmon are at last getting their acts together'.

If we fail, the fate of the salmon has far wider implications than the survival of fish and the cleanliness of our rivers. The truth of the matter was put most clearly by the writer Roderick Haig-Brown: 'The salmon runs are a visible symbol of life, death and regeneration, plain for all to see and share . . . If there is ever a time when the salmon no longer returns, man will know he has failed again and moved one step nearer to his own final disappearance.'

Chapter 10

A Gaggle of Geese

Of all the country sports in Britain, wildfowling has a special niche. To follow their sport wildfowlers trudge to some of the most isolated and inhospitable places in the country; remote estuaries, mud-flats and marshes, where the cold wind whistles in from the sea and where the memory of street lights, central heating and parking meters seems like part of another, unreal world. The tradition of wildfowling has a long and honourable history. In the East Anglian Fens it provided work for professional wildfowlers, who were as much a part of the fenland scene as eel fishermen and fen skating. They would catch their quarry – wildfowl and plovers – by a variety of means: decoys, nets, traps, snares, shot guns and punt guns. They were reaping part of an abundant natural harvest which provided food locally, with the surplus sent for sale in the towns. The 'fowlers themselves were hardy men; they had to be, to face the discomforts of cold and damp, as well as the dangers of rising floods and racing currents. Some of them, for warmth, would wear goose grease and brown paper over their chests – applied in the autumn and removed the following spring. If the wearer had an efficient nose, then other considerations came into effect, as the vest was considered to be 'wearable until the smell becomes unbearable'. Due to the nature

The punt gunner and wildfowler were and still are the hardest breed of country sportsmen.

of their task the 'fowlers became excellent naturalists, understanding their quarries and the natural influences on them – the rain, winds and changes in temperature. They became familiar with all the inhabitants of the marsh and foreshore, and expertise in weather and wildlife lore grew with experience and helped them with their work.

Today's 'fowler still has to be hardy, as he goes with windproof jacket, waders and eager gundog (to retrieve any shot birds) – marching off into the muddy, watery wilderness. The 'softer' ones wear gloves and balaclava helmets to protect hands and heads in cold weather, but I have seen others, in freezing conditions, with red, unprotected fingers and ears, seemingly oblivious to the cold. They are a complete mixture of individuals; because it is sometimes hard, uncomfortable shooting, it is accessible to a wide range of people, and so it has a large 'working class' following. It includes ordinary country men – cowmen or coalmen – continuing old family traditions, as well as many from the towns, lorry-drivers and labourers, who find that wildfowling enables them to get out into real country and gives them a taste of adventure. The same feelings are experienced by some professional people, solicitors, accountants and school teachers who come to prefer the saltings and water meadows to their work behind office desks. There are sportsmen, too, who have the whole range of shooting at their disposal – grouse, driven partridges, and high flying pheasants – yet they prefer wildfowling. They are attracted by its isolation and challenge, and they find the lure of wild places and the calls of ducks and geese irresistable. To wildfowlers, companionship is also important; a shared experience with a friend, or simply the presence of a faithful dog.

The moods and attractions of wildfowling have been described many times in books, the most famous coming from the redoubtable pen of Col. Peter Hawker. More recently Denys Watkins-Pitchford, better known as 'BB' – with two illustrations in this book – has caught the atmosphere exactly. In the *Dark Estuary* he recalls the kind of evening that will have been experienced by many wildfowlers:

> I could watch for ever the coming of the tide on a quiet winter's night of moon and stars. It is a scene which must be dear to the heart of every wildfowler.
>
> Soon there is water sucking and gulping obscenely among the hollows of the 'brew'. It comes washing in round your waders and retreats again, gathering itself for another onslaught.
>
> And everywhere are birds, birds – the night is full of wings and cryings. Curlews, redshanks, dunlins, 'oysters' – they are all in the air at once. It is amazing where they have all come from. The answer, of course, is that they were out on the tide edge and have retreated before it to the shore.
>
> And then comes the moment that you have been waiting for. Above the bustle and roar of the breaking tide, above the yodelling and the pipes, the reedy cheeps, and harsh alarm notes of the 'shank, there comes the yelping of the barnacle geese.

Morecambe Bay – a paradise for wildfowl, waders and wildfowlers alike.

Familiar birds of the foreshore. LEFT: *The curlew.* RIGHT: *The redshank.*

> I do not think that the cackling of greylags or the chorus of the pinks is quite so moving as the sound of a barnacle pack in full cry. It is unearthly. Phantom hounds most certainly and Herne the Hunter should be riding after!

Barnacle geese are not protected, but 'BB' portrays both the mood, and the respect even affection, the true wildflowler feels for his quarry.

Estuaries, marshes and water meadows are among our most threatened habitats and in most parts of the country there are wetlands that have been saved partly because of their wildfowling connections. Many are still used for shooting during the winter season, with the local wildfowlers often taking part in the management and wardening of their areas throughout the year. Some places, once famous for wildfowling, have also been purchased and turned into wildlife reserves, bought by various conservation bodies – the RSPB, English Nature or local wildlife trusts. Even then the wildfowling rights are sometimes offered to the wildfowlers, for the ducks and geese they take, can still be seen as part of the natural harvestable surplus. The end product, roast wigeon, mallard, or even goose, with peas, sausages, apple sauce, bread sauce, roast potatoes, sage and onion stuffing, and good gravy, preceded of course by light pudding, is a meal that I would never refuse.

The best known area where conservation and wildfowling takes place side by side is the 6000 acres of the Ouse Washes in Cambridgeshire. The Washes were formed to help drain the Fens in the seventeenth century; then 'marsh shepherds' grazed their cattle on the lush summer grass, and in winter, with water and flooding, huge flocks of wintering wildfowl transformed them into a paradise for the wildfowler. Today wigeon, teal, pochard and tufted ducks still arrive in large numbers, while the Washes have become internationally famous for their winter populations of Bewick's and Whooper swans.

Another area to owe its wildlife importance to wildfowling is Old Hall Marshes in Essex, one of the RSPB's most prized acquisitions. The former owner was a keen wildfowler and so traditional coastal livestock husbandry suited him, for the grassland and water attracted large numbers of wildfowl during the winter.

Old Hall Marshes is an interesting place – 1500 acres of ancient Essex grazing marsh on the Blackwater estuary. At one time many of the farms along the low lying Essex coastline were similar; now it is almost the only one left. It is an ideal mixture – 120 acres of saltings, 170 acres of improved grassland, 655 acres of rough grazing marsh and 189 acres of water, including an old commercial decoy.

In the summer it has a lazy, pastoral feel, with lark-song and grazing cattle. The rough grassland is an astonishing sight, with thousands and thousands of ant hills. The mixture of grass and water means that there are many species of breeding bird; lapwing, oystercatcher, ring plover, redshank, shelduck, mute swan, Canada goose, mallard, shoveler, tufted duck, pochard, moorhen, coot, water rail, little grebe, cuckoo, reed and sedge warblers, meadow pipit, skylark and reed bunting. In the few low thorn trees, artificial herons' nests have been built, to try to attract herons in to breed, and over the sea wall on a small island, terns often lay their eggs. Invariably the nests get swamped and destroyed during the high spring tides – giving them the local name of 'bird tides'.

Just inside the sea wall is a long dyke – the 'borrow dyke', so named because the earth from it was 'borrowed' to build the sea defences.

It contains fresh water, and the terns often fish along it. There are other areas of freshwater too, including the old duck decoy surrounded by reeds. The reserve is not limited to birds alone and has rabbits, hares and foxes. Over recent years an experiment has been taking place concerning the value of predator control. Four years of no predator control has been followed by four years of predator control – to see if the number of breeding birds is affected. So far the results have not been assessed, or been made public.

The importance of Old Hall Marshes increases in the autumn, for then the wintering wildfowl arrive, including up to 4500 wigeon, 3500 brent geese, 600 shelducks, 1000 teal, 300 mallard, 300 goldeneye and 120 pochard. Numbers of wigeon (6000) Brent geese (4500) and shelduck (1200) show a marked decrease in sixteen years. Whether this indicates a change in habits, population or reserve management is not clear.

My winter visit was on a perfect winter day of open sky. The sun illuminated the tassle-headed beds of reeds (Phragmites), and there was a large flock of grazing Brent geese. The Brent goose is an attractive bird: I have seen flocks arriving over the North Sea in the autumn, during appalling weather, and feeding on winter wheat near the Wash. At Old Hall Marshes some were feeding, while others were flying in scattered skeins, calling conversationally as they flew from the mudflats to the grassland.

The reserve acts as a refuge for the geese as it draws them away from local farmland. When the geese do go onto the neighbouring arable fields, the farmers can claim compensation for any damage to winter crops. At one time Brent numbers fell to danger level, and they have been protected since 1954. Since then numbers have steadily risen, however, and some wildfowlers are asking for the goose to be returned to the 'quarry list'.

The increase in the number of Brent geese is one of conservation's success stories. The East Anglian Brent geese are beautiful birds and are in fact the dark-bellied sub species.

In the last twenty years the wintering population has increased by 80% to just over 90,000. Part of the success is undoubtedly due to the fact that the geese have changed their eating habits. At one time it was thought that Brent geese would only eat eel grass – to be found in diminishing amounts among salt marshes. Suddenly however – they discovered the attraction of the shoots of winter sown wheat and as a result their diet, as well as their winter condition has changed dramatically.

Personally I could never shoot a wild goose, but I will readily eat them as they are delicious. In view of the increase in numbers it seems to me that the request of the wildfowlers for the Brent goose to be returned to the quarry list is entirely justified. But totally illogically it seems that the change of status will not be granted. Once a bird or animal is 'protected' its protection remains – regardless of whether its circumstances improve. This often means conflict, with concern over high numbers caused by genuine farming damage being almost totally ignored by the conservation lobby and its tame MPs. As a consequence sportsman are then reluctant to stop shooting other species whose numbers are in decline, for fear that any ban would last permanently. At the moment I believe that a strong case can be made against shooting snipe and golden plover because of falling numbers – but the shooting lobby is reluctant to suggest this because shooters have seen what happens with 'protection' in the case of the Brent goose, the hen harrier and the sparrowhawk.

What British conservationists have forgotten, unlike those of many other countries, is that population management is just as important to wildlife as habitat management. In an overcrowded world in which 'natural balances' are no longer possible because of the activities of man – man often has to intervene to mimic that balance. Consequently when populations of some species fall to danger level – they need help and protection to recover. When populations explode – then taking a surplus – a natural harvest, or controlling numbers to protect more fragile species is good conservation based on common sense. Consequently it was right to give the Brent goose full protection when numbers fell dangerously – it is ridiculous to keep that protection when some conservationists believe that the present breeding grounds have become overcrowded.

Similarly when sparrowhawk numbers plummeted through DDT and Dieldrin poisoning it was right to give it complete protection. Now, when sparrowhawk numbers could actually be damaging the survival of other birds – such as the breeding lapwing in East Anglia – their protection is absurd and based on 'political correctness' and not conservation.

On the day of my visit it was clear what the foxes had been eating, as there were several patches of mallard feathers, showing where 'dinner' had been taken. Ducks were not the only item on the menu, for among the ant hills were a number of places where attempts had been made to dig out voles.

Not all the remaining areas of wetland in Britain are large or part of protected nature reserves. In total contrast to the Ouse Washes and Old Hall Marshes I came upon another superb place for wildfowl on a much smaller scale, in Dorset. There, the farmer keeps an old system for flooding water meadows in working order, specifically to attract wildfowl and snipe. The farm itself is situated near Puddletown in the heart of Hardy country, and the meadows adjoin a small chalk stream, a tributary of the river Piddle.

In winter a number of sluices are used to raise the level of the stream. The water then spills over into a network of channels and ditches that flow parallel to the main stream, but several feet above it. The ditches were dug out well over a hundred years ago, to Dutch design, to help get better spring grazing. Now Hugo Wood-Homer uses it to bring in wildfowl and waders, particularly wintering wigeon, At first the main attraction was

Bewick's swans and other wildfowl in the freezing Ouse Washes.

shooting, particularly snipe and ducks, but Hugo has become so attached to his wildlife that he has stopped shooting. His son-in-law now runs a small pheasant shoot on the farm – but the water meadows are out of bounds and have become a sanctuary. He is not against shooting – but it is no longer for him. When I walked by the stream on a watery February day, there were large flocks of wigeon together with mallard, and teal. There were lapwings, too, as well as snipe and one mute swan. Coastal waders such as dunlin and redshank like to visit, as well as numerous other attractive birds including ruff and black-tailed godwits. Green sandpipers also pass through. During the winter, peregrines, hen harriers and merlins hunt the water meadows and Hugo Wood-Homer has seen a merlin take a snipe in mid-air. Hares are numerous, and one sprinted off, hurdling expertly over the full drainage ditches as it went. The farm is 1300 acres in size two-thirds of which are arable, the rest being grass for sheep and dairy cattle. It was originally owned by Hugo's grandfather, who was something of an eccentric. He sold the Elizabethan manor house which went with the land and built a Victorian folly. The 'farmhouse' now looks like a cross between a castle and a church, complete with stained glass windows.

Like so many others, through shooting, Hugo Wood-Homer's interest has become far wider: from believing that game and good farming are completely compatible – wildlife for its own sake is now a major influence. He has hedges, 'weeds' along the edges of his farm roads, copses, two lakes and a reedbed that he has planted. In addition he has twenty acres of unimproved downland on a south facing escarpment, complete with an old hazel coppice from which hurdles were once made. Spindle, dogwood and guelder rose all grow well, as do cowslips which provide a springtime carpet. The butterflies of chalkland are well represented with the marbled white and three blues – the small, common and chalkhill. Holly blues are also to be found on the farm, in addition he has recorded 750 species of moth. Tawny owls and sparrowhawks breed and hobbys have nested successfully. Whimbrel, bittern, osprey, black tern, great grey shrike, wryneck and ring ouzel are among other birds to have visited.

The free flow of water has made it possible to create lakes for wildfowl and fishing, and already mallards, little grebes, coot, tufted duck and Canada geese are breeding. But his enthusiasm is not linked to large birds, or rare birds only, and in the summer of 1986 he recorded the number of pairs of warblers on the farm. He counted 32 pairs of black-caps, 17 sedge warblers, 13 whitethroats, 15 chiff chaffs, 5 garden warblers, 18 willow warblers, 1 reed warbler and 1 lesser whitethroat. He also gets the occasional grass-hopper warbler.

It is astonishing how an interest in wildflowling has triggered a wider interest in overall conservation with many people. The most obvious example is that of the late Sir Peter Scott, the artist, naturalist and founder of The Wildfowl and Wetland Trust. Another is the country writer, shooter, journalist, raconteur and ex-teacher John Humphreys. Almost twenty years ago he bought a 'boring' piece of Cambridgeshire Fen close to the River Cam. It was seven and half acres of nothing – anybody who knows the Fens will know exactly what seven and half acres of nothing looks like – flat, featureless and on a gloomy winter's day, bleak.

Now both John Humphreys and Hunter's Fen have been transformed. The man took early retirement to take up writing, shooting and fishing – and the Fen has been turned into a piece of wildlife rich wilderness, because of the writing, shooting and fishing of John Humphreys.

Through a mixture of hard work and vision the seven and half acres now have one pond, of an acre, for fishing and a smaller pond, a third of an acre in size, where the fish are deliberately kept out for the sake of wildlife – frogs, toads, tadpoles and the larvae of assorted creatures that in summer reflect sunlight form their blazing wings. Now, in high summer the paths between the trees, shrubs , weeds, reeds and wildflowers give a misleading feel of distance – seven and half acres? – it feels like fifty – winding through willow, thorn and scrub. Reed warblers sing, damsel and dragonflies dart: 'Look', says John Humphreys with a mixture of pride and wonder, 'a kingfisher'. Still after twenty years the sight of a kingfisher stops him in his tracks. He remembers where each tree came from, when it was planted and why. He looks at the ponds and the planted reedbeds and remembers flat Fen – what a wildlife and wilderness achievement – what an example of bleak, drained, 'improved' Fen being brought back to life. And what was the driving force? Shooting a few wild ducks over water – that old tradition of the Fens. Now that would-be wildfowler shoots 100 wild cock pheasant and 150 ducks 'with restraint' over the course of a season and Hunter's Fen should perhaps be re-named Wildlife Fen. On name changes perhaps the RSPB should change its name from 'The Royal Society for the Protection of Bird's to' The Royal Society for the Conservation of Birds'. Such a change might bring in more common sense.

Bill Makins of Norfolk is yet one more duck shooter turned conservationist. Again he started by shooting a few ducks over ponds in a flooded meadow. Then his interest went beyond the skill of concealment, shooting and retrieving to understanding the life cycle of the wildfowl themselves. With the help of gravel digging his flight pond turned into sixty acres of pools and gravel pits with 200 acres of meadowland.

The pits became home for many of Britain's most attractive ducks as The Pensthorpe Waterfowl Park was created. Pinioned and captive breeding birds attracted wild birds too – pochard, gadwall and teal. Gradually Bill's captive breeding programme spread to avocets, cranes and corncrakes. One man's quest for a wild duck on the dinner table turned into one of the most impressive wildfowl conservation projects in the country.

Men like Hugo Wood-Homer, John Humphreys and Bill Makins are not alone as many wildfowling clubs have become important conservation bodies in their own right, with the interests of their members stretching far beyond shooting for sport. Kent wild-fowlers have a number of conservation projects in hand, and the club now calls itself the Kent Wildfowling and Conservation Association. Similarly the Chichester Harbour

Wildfowlers' Association has been involved in a number of schemes of real conservation merit.

In addition to playing an important part in managing and creating good wildlife habitat, wildfowlers have one other important function. The British Association for Shooting & Conservation (BASC) has a membership of over 100,000 including 1575 affiliated clubs, of which about 200 are wildfowling clubs. As a direct result they form an influential pressure group, which could become more important in the future. Estuaries are a favourite haunt of wildfowlers, yet many of them are currently under threat. The RSPB has listed no fewer than 30 important sites for wintering wildfowl and waders, which are being investigated for tidal barrages, industrial development, ports, marinas and other recreational facilities. A total of over 800,000 birds make use of these areas for feeding or roosting. Among the places being considered are the Severn, the Mersey, Morecambe Bay, the Solway and the Wash, all of which are of international importance. Morecambe Bay has been suggested as a site for a tidal barrage, yet its wintering bird population can rise to over 246,000. Similarly the Wash has 286,000 and the Solway 128,000. The Severn, too, is a favourite of those who advocate tidal power, yet the estuary's mud flats attract many thousands of wildfowl and waders. In addition tidal barrages could interfere with the salmon wanting to 'run' up river to spawn, and the famous Severn bore would be no more. I can never understand those 'green' politicians and pressure groups who claim that tidal power in estuaries is a viable form of alternative energy. The damage it would do to wildlife would be even greater than the methods they already condemn. Because of this threat, the larger the number of opponents lining up against the Department of Energy the better, and so it is important for wildfowlers to be alongside the RSPB, English Nature, and the various other wildlife and environmental organisations.

Snipe (by A. Thorburn) love flooded water meadows and marshy areas.

But although wildfowling has many good and genuine links with conservation, it has to be said that there are some difficulties, both ethical and ecological. One of the problems is the fact that over half of those who actually shoot wildfowl do not belong to BASC, and so they do not always reflect its standards or adhere to its code of practice. I believe it is wrong for any wildfowler to go shooting without a dog to retrieve wounded birds, yet it is easy to see wildfowlers with no dogs, and many of those visiting the Ouse Washes have no dog.

Another problem involves the 'marsh cowboys' – whose behaviour causes critical letters in the sporting press every year. They are shooters who quite literally shoot at anything that moves, in range or out, sometimes even using illegally semi-automatic weapons (illegally for wildfowling, that is). The problem is at its worst in Scotland, where, once more, open access to the foreshore means that wildfowling at places such as the northern side of the Solway can be uncontrolled. Hoteliers cash in with arranged shooting weekends for all and sundry, including foreigners, when shooting standards can be even worse. On the southern side of the Solway, in England, where the shooting rights are owned by the Crown Commissioners and leased to clubs, there is no such chaos and mayhem.

Even on SSSIs such as the Ouse Washes there are some privately owned Washes, not leased by responsible clubs, where anybody with a shotgun certificate can purchase a 'day ticket' and shoot. The shooter paying for the ticket may not have been wildfowling before and his actual knowledge of wildfowl can be virtually nil. Yet he can still go shooting wildfowl and waders – in an area which includes both legitimate quarry and protected species. When I asked an acquaintance if marsh cowboys still visited the Washes he said that things had improved slightly over recent years: 'But there are complete idiots buying day tickets and shooting at anything.' This was confirmed by Cliff Carson, the RSPB warden on the Washes. When I wrote *The Hunter and the Hunted* twenty-six years ago, Jeremy Sorensen, the then warden, considered that a flock of Bewick's swans could not fly up the Washes without being shot at. Cliff Carson reports that this is now virtually unknown. His main concern is those washes where day tickets are sold, or where responsible wildfowling clubs are not involved. By one such wash in the 1986/87 season a ruff (an attractive, protected wading bird) was found shot dead. Similarly, the warden on the Wildfowl Trust's reserve at Welney has grave reservations about day tickets: 'Often the people who shoot with a day ticket are "townies" with no knowledge of wildfowl. I have nothing against the genuine wildfowler who belongs to a good club, but if there are no ducks about, the day-ticket shooters tend to let fly at anything. I know in recent years several cormorants have been shot. Each year, too, four or five whooper and Bewick's swans stay behind, because shot-gun damage prevents them from flying back to their breeding grounds with the rest of the birds.'

To me the solution seems to be that the selling of day tickets should be banned, unless undertaken by responsible clubs. Also, any new gun owner wanting to move into wildfowling should be required to undergo an identification test for wildfowl and waders. Already such tests are compulsory in some European countries. It could be administered by English Nature with advice from the RSPB and the BASC. Some responsible clubs already make their own efforts to help the new wildfowler become competent. The Ely and District Wildfowlers have an arrangement with the Wildfowl Trust to visit the Welney reserve, to see the various species, with an expert on hand to help with identification.

Another main problem again involves lead. Quite simply, in some places cartridges used for shooting wildfowl contain lead. As a result, just like the old lead weights of fishermen, a high level of shooting can cause a potential hazard for a variety of dabbling and diving ducks, as well as swans.

Whenever water or wetland are heavily shot over, there is a risk. If the water level is lowered, bringing much new mud to within dabbling depth, then problems can occur. The worst case of such poisoning in Britain occurred between November 14th 1986 and December 20th 1986 at Loch Spynie in Morayshire. There the water of the loch was accidentally lowered, exposing much more of the loch's sediment – including lead pellets, as it was once regularly used for wildfowling. As a result 285 dead greylag geese were found and it is estimated that up to 500 could have died altogether. From the corpses it was discoverd that 86% had died from lead poisoning caused by ingested lead gunshot pellets, and 14% had died from lead poisoning from anglers' lead weights. A few years ago, 50 Canada geese died in another probable lead poisoning incident, at a brick pit in Bedfordshire used by a gun club.

In America several such 'die offs' have been experienced with the result that in 1991 all wildfowlers were required to use cartridges containing non-toxic shot. The first attempts to produce steel shot ended in failure with the hard shot damaging barrels and the cartridges themselves not comparing ballistically with lead. Now, however, the second generation of steel and tungsten cartridges have overcome many of the problems. In fact some American 'hunters' actually favour them, wearing lapel badges proclaiming: 'You've got to have steel balls to shoot ducks.'

Lead shot for wildfowl was actually banned in England from September 1st 1999 – but lead shot is still legal for wildfowl in Wales, Scotland and Northern Ireland.

The legislation, as usual, was ill-thought out. Most shooters still prefer lead shot – as ballistically lead remains superior to the non-toxic alternatives – but the government of the day made a blanket ban on shooting all wildfowl with lead – rather than a selective ban on the special areas where wildfowl are shot. Special areas of wetland such as the Ouse Washes need protection on environmental grounds – but the legislation even prevents lead being used for shooting geese landing on a potato field or mallards flying to feed on stubble after harvest. So – a pheasant can be shot over stubble with lead – a mallard duck cannot – 'bonkers' is rather a flattering assessment of the situation.

There is one abuse too – which can be mistaken for 'wildfowling' but which in reality has nothing to do with traditional wildfowlers – although to the uninformed it can give wildfowling a bad name. On some commercial shoots, mallard ducks are reared and released in very large numbers – in a similar way to pheasants. The big difference being that the ducks will stay on or near their rearing pond or lake as 'home'. Even once shooting starts they will simply circle over head; the result is large numbers of dead and dying ducks. In my view this type of 'duck shooting' can hardly be called sport, as there is little skill or challenge demanded ; it is certainly not 'wildflowling'.

But despite this and the excesses of the marsh cowboys, the overall influence of the wildfowler on wildfowl and conservation is extremely positive. Just as this book is called with justification, *The Fox and the Orchid* – so this chapter could have been called 'The Mallard and the Marsh'.

Geese circling – the spirit of wildfowling from 'Dark Estuary' by BB.

Epilogue

Relooking at the impact on conservation of hunting, shooting and fishing after a gap of sixteen years has been a fascinating challenge. In that time the pressures on land and the general countryside from housing development, new roads, new airports and increased leisure time have grown considerably. On top of this, because of the collapse in agricultural prices, most farmers have had to work their land harder – more intensively – to survive. The old throw away lines, still being thrown away of: 'Oh, I've never seen a poor farmer' and 'you never see a farmer on a bike', have never been less appropriate. The tragic truth is that the average farm income today is just £16,000 per year and in 2002 it was a mere £10,000 and so there are many farmers in a state of desperation, who can hardly afford a new bike. As a consequence small farms and family farms are dying and being consumed by the large – big farms are getting bigger and their methods are becoming ever more industrial. Over huge swathes of land, covering both cereal and livestock farming, industrial farms are becoming what the government wants them to become – 'efficient'. 'Agri business' has replaced 'agriculture'; the 'land manager' has replaced the 'farmer' and wildlife has been decimated as populations of numerous birds and animals continue to fall.

For those who doubt these straight forward facts, ask yourself three simple questions – when did you last hear the cuckoo? When did you last have a swallow nest in your garage and when did you last watch hares in their March madness in the early spring? At the same time when did you last travel through the countryside and see the tell-tale sign 'Farm For Sale' and the rows of implements and items of a lifetime of farming being sold under the auctioneer's hammer?

I have been beyond the 'For Sale' signs: I have had broken men and women crying on my shoulders because they can no longer cope as an uncaring government makes no effort to listen or to understand. I have seen fleeces smouldering because they are cheaper to burn than to sell. Like many too I have been to a funeral when all hope seemed to have gone for a working, living, happy countryside.

In this atmosphere and period of change it is still the hunting, shooting and fishing farms that stand out almost like islands of green sanity surrounded by factory farming – places where trees, woods and rough places left for fox and pheasant provide ideal conditions for so much more. What they provide goes beyond habitat for wildlife as the seasonal calendars of hunting and shooting provide social connections for rural communities under pressure. Gamekeepers, shooters, beaters, kennelmen, puppy walkers, hunters, car followers and many more find themselves with occupations and celebrations – seasonal events bringing them together and giving them points of contact to combat the air of isolation generated by economic depression and political attack.

Astonishingly as I have been revisiting this book so these political attacks have become shriller and the threats to hunting and shooting in particular have become greater, despite their importance to the conservation and social cohesion of the traditional countryside. Why should this be when there are so many other political issues that logically should take precedence? And why pick on environmentally friendly activities when there are real issues of pollution that appear to cause our politicians no concern at all? At the end of the *The Hunting Gene* I quoted one of my own poems:

The rain falls,
The spring rises,
The stream flows
The river winds,
The sewage seeps,
The acid burns.
The poison bubbles,
The fish die
The trees sigh.
The dolphins weep,
The gulls cry,
The rain falls;
The cycle of despair.

This poem still contains the real issues – the issues that short term, ego-tripping politicians will not address.

By coincidence, as I have been updating *The Fox and the Orchid* scientific research from the University of Kent has confirmed the importance of hunting and shooting to conservation. The Report states that landowners involved with hunting and shooting retain 7.2% of their land as woodland cover, compared to only 0.6% among other landowners. Thomasina Oldfield, one of the researchers said: 'The study suggests that landowners may conserve important habitats if they are involved in hunting and shooting. Therefore our study implies that the role of landowners in voluntary habitat conservation should also be considered in current debates on field sports.'

But despite the many positive sides to country sports, not all the activities connected with hunting, shooting and fishing are without blemish, but the plusses far outweigh the minuses. That is why I believe 'prohibition' concerning country sports is unacceptable on the grounds of freedom of the individual, democracy and animal welfare. But despite this some regulation is required to ensure that the sports are carried out responsibly.

Coarse fishing still seems to be the sport that raises most concern and it is a reflection of the lack of integrity of our politicians that in real political terms angling is the most secure of all country sports. The closed season continues to be too short and the starting date of June 16th in my view should be put back until August 1st. It is almost laughable how the opponents of hunting with hounds criticise the 'disturbance' caused by mink hounds as they pass along a river. But that is exactly what they do – pass along – the disturbance is minimal and short lived – whereas the disturbance during peak nesting time for waterside birds from fishing can last for hours on end and large areas of reed banks (including nests) are often flattened as fishermen seek to find the best fishing position. Even worse is the fact that still water – ponds, lakes and canals can have no close season at all, at the discretion of the owner.

I know of one 'fishery' which in effect is simply a small artificial pond with no conservation value whatsoever. In early May a van arrives and fish are poured into the pond from a container; the next day fishermen arrive to fish for these reared, trapped creatures. Enough fuss is made about 'reared pheasants' – at least they become free – what about 'reared', incarcerated fish? Imagine the outcry if foxes or hares were hand-reared and then released into a field from which they could not escape. Amazingly some opponents of 'blood sports' actually believe that this happens with hare coursing – basing their objections on facts is just too much for them.

Incredibly too live-baiting still occurs in angling – when a live fish is turned into a 'bait' with hooks threaded into its body to attract pike or some other predatory fish. Yet another recent report has confirmed that fish do feel pain; how this practice manages to avoid the attention of the animal welfare lobby or the various fishing bodies themselves is a mystery.

Shooting too has its blind spots and the excesses and abuses of the few manage to tarnish the whole sport. On John Wilson's farm in Suffolk all the pheasants are completely wild and a good day will yield sixty pheasants. Gamekeepers, shooters, beaters and the retrieving dogs will have enjoyed a good, unpredictable day, and all the shot pheasants will end up on the dinner table. At the other, unacceptable end of shooting, thousands of pheasants are reared and released, with some shoots releasing pheasants at intervals during the season to ensure a continuous supply. 'Bought' days can result in 800 or even 1000 pheasants being shot – such shooting is not 'hunting' – it is simply target shooting – using live targets. Perhaps those who indulge should be persuaded to take up clay pigeon shooting and shoot at lumps of clay instead of real birds. To make matters worse, on some of these 'big pheasant days' there are simply too many birds for game dealers or butchers to deal with and so a hole is dug and the dead pheasants buried. My disgust at these shoots is total and the damage they do to the reputation of the countryside is enormous. In these instances shooting is seen just as a business, or a status symbol, and no respect is given to the quarry or to the countryside.

I know of one French partridge shoot in East Anglia where the dead birds are buried anyway – the question of a game dealer not coping with numbers does not even arise – burying is the policy of the shoot. The guns arrive – the money is taken and the live birds/dead targets are shot and buried. To me justifiable country sports are founded on killing to eat – and killing to control populations – burying pheasants and coarse fishing – catching a fish and putting it back, satisfy neither criteria.

Unfortunately some shooters seem driven by the need to shoot. A few years ago I presented a relaxed country programme on Anglia Television called *Countrywide.* Early one September we filmed a partridge shoot in East Anglia. There, in front of the camera one of the guns who had few partridges flying over his position started firing at pheasant poults (young pheasants) weeks before the pheasant shooting season had started. Consequently 'regulation' is required not 'elimination' by an Act of Parliament – after all football will never be banned because of foul or violent play – so why should country sports be threatened because of the activities of a small minority?

With fox and deer hunting the main concern comes from the lack of thought from some huntsmen. Britain is a very over-crowded island and if a fox or deer decides to head for a village centre it seems the height of madness to continue the hunt. There are people with genuine objections to hunting and a fox being killed outside their French windows as they watch Saturday afternoon *Grandstand* on television is something to which they should not be subjected. Consequently some huntsmen and Masters should be more aware of the feelings of others and stop hunts if the quarry heads into suburbia. Such action is not 'yielding to anti-pressure' – it is behaving properly and respecting the views of others. In a free society there should be a 'right' to hunt – but there should also be a 'right' not to have the hunt on your property if you do not want it.

But inevitably the one word that dominates all comments on country sports is not 'conservation', 'freedom' or 'culture' it is 'cruelty'. Forget broiler fowl – the concentration camps of cheap food policies – (for a description of broiler fowl death read *The Hunting Gene,* Chapter 7), forget the closure of slaughter houses – Britain with a population eight times greater, with a land mass six times larger, has fewer slaughter houses than Switzerland; forget the murderous halal slaughter – recently described as cruel by The Farm Animal Welfare Council. Forget the nine million cats killing 275 million birds and animals a year; forget the wildlife wipe-out caused by development and forget the destruction of Brazil's forests so that soya bean monoculture can replace the world's oxygen factory – allowing affluent, western vegetarians to feel morally superior. No – to suit inflated egos and suspect morality, society and Parliament do not regard these activities as cruel – it is only HUNTING that is cruel – hunting, when the actual death is almost instant.

Society tolerates far more 'cruelty' in other aspects of twenty-first century living but for the sake of political correctness it is hunting that is always pulled out of the hat first. It is an extraordinary state of affairs for in my view hunting causes less distress than the other sports of shooting and fishing and what about broiler fowl and halal slaughter? It should also be remembered that in the year 2001 it was the fox loving government that quite ruthlessly, wantonly and cruelly killed 10 million animals during foot and mouth. Ten million animals that could have been saved by vaccination and 10 million animals, many of which, I believe, were killed quite illegally – a fact confirmed by Professors David Campbell and Robert Lee of Cardiff University. This means that many of the politicians attacking country sports, do so with blood already dripping from their own hands.

One simple guide on the cruelty involved in sport is to compare wounding rates. What are the wounding rates of hunting, shooting and fishing? The answers are easy and startling. The wounding rate in hunting is 0% – the pursued animal is either killed or it gets away. The wounding rate in pheasant shooting is anything between 25% and 75% depending on the competence of the shooters. The wounding rate in fishing is 100% as every caught fish is hooked and damaged. In fact the figure is more than 100%, as many fish break free – still with hooks, spinners and line attached.

The problem is of course that the whole question of cruelty – especially the perceived cruelty linked to hunting is not based on fact – it is based on political opportunism, expediency, self promotion and ignorance. Put even more simply much of the 'moral argument' against hunting is based on lies – as the late Malcolm Muggeridge wrote in *The Green Stick (1972);* 'People, after all, believe lies, not because they are plausibly presented, but because they want to believe them' – that in a nutshell sums up much of the opposition to hunting.

As I write – hunting has just been theoretically banned by the House of Commons. But the whole Parliamentary process has been based on lies. When the government commissioned the Burns Report it was done ostensibly to help Parliament make an informed decision – yet Parliament almost completely ignored Burns.

Immediately after Burns was published Alun Michael – an anti-hunter put in charge of a 'neutral' anti-hunting Bill suddenly announced new criteria, – 'utility and cruelty', was hunting useful and was it cruel? If the same measurements were applied to shooting, fishing, show jumping, horse racing, keeping pets, cats, hamsters, goldfish – all would be banned – but no, **hunting** was, and is, the only activity to be measured using these two bizarre words, plucked out of the air by a jobsworth junior Minister. To help the government formulate its policy Mr Michael – described by many as a 'Blair Poodle' – then asked various organisations and individuals to appear before him to give evidence. As Chairman of the CRT and author of three books on country sports I asked to be called – I was **not** invited. I sent a copy of my book *The Hunting Gene* to Mr Michael – **he sent it back** – what an open mind !

So much for his desire to receive evidence. Once in Parliament the whole process was seen for what it was – the 'debate' was simply an excuse for the old 'class warriors' to settle old scores and the Prime Minister – sinking deeper into trouble among his own back benchers had a bone he could throw them – indifferent to the needs, feelings and realities of real people in the countryside. Parliament was seen for what it was and is – arrogant and out of touch – the £1,000,000 given to the Labour Party in 1997 was being earned – and how much more, if any, has been promised for the future? Tony Blair cried crocodile tears for minorities in Iraq; yet he has no regard for his own traditional rural minority in Britain.

So urban domination was seen to replace democracy, and fixed political positions based on vindictiveness and prejudice, replaced honest debate. As Laurens van der Post wrote in *A Walk With a White Bushman*: 'Western Europe was once governed by a country mind – even city states had a keen country awareness – but the country mind has greatly diminished. People who know nothing about the land suddenly sit and rule like gods over the fate of other men'.

Some people have been surprised at the lies spoken in Parliament about hunting; their naivety is touching as the government of Tony Blair has a history of bluster, blame, sham and dishonesty – look at the lies behind the policies for Serbia, Kosovo, Afghanistan, Iraq, foot and mouth, hospitals, schools etc. Among them all are the threads of deceit and dishonesty being spun out of control. For anybody doubting the deceit of the government, look no further than farming, the Burns Report and foot and mouth. When Sir Donald Curry was asked to produce a Report on the future of Farming and Food in the aftermath of foot and mouth, the causal observer, prompted by the BBC, could have been forgiven for thinking that a serious effort was being made to address farming's problems – dream on. Potentially three of the most damaging forces affecting Britain's family farms are the CAP (Common Agricultural Policy), globalisation, and the expansion of the EU. But there in the terms of reference for Sir Donald was the instruction that his Report has to be 'consistent with the government's aim for CAP reforms, enlargement of the EU and increased trade liberalisation' (which means globalisation).

In such limited circumstances it does Sir Donald Curry no credit, it seems to me, that he took on the truncated and deliberately misleading task. It is odd how bad governments, bad politicians, bad people, rich people, crooks, gangsters, and fraudsters always manage to persuade a mixture of good people, gullible people, stupid people, greedy people, conniving people, jobsworths, quislings and buffoons to help them with their devious ways. Similarly Lord Burns and his team – the people commissioned to study hunting and all its implications were then barred by their terms of reference from making the two most basic judgements – judgements needed to be made by informed people – was hunting cruel and should it be banned? No, let the uninformed in the House of Commons make those judgements.

Then of course there was foot and mouth. There are so many stories concerning the horrific foot and mouth outbreak in 2001 – the arrival of the virus and the actions, legal and illegal of the government that only a full-scale Public Enquiry will yield the sorry and appalling truth. The government will not allow that Public Enquiry to take place, in case it reveals a web of deceit , half truths and total lies.

It should therefore surprise no-one that the same standards and lack of integrity should be used in the hunting debate – which will inevitably move on to shooting. What Tony Blair and his henchmen have been creating is the cultural cleansing of the British countryside. In a so-called multi-cultural society Britain's indigenous rural culture has experienced an unprecedented political attack. In Britain in the year 2000 1.7% of the population was Afro-Caribbean and 0.5% was African. In crude terms this means that 2.2% of the population was black and to attack the culture of black people is rightly classified as racism. Consequently by attacking rural culture (10% of Britain's population is considered 'rural') I believe that Britain's urban Parliament has shown itself to be fundamentally and institutionally racist.

After hunting – despite the political assurances to the contrary – shooting will almost certainly come under attack – attacked by people separated from nature and the land.

Country sports have over the years helped to shape and preserve the countryside and the wildlife that depends on it. But now they have come to symbolise something even more valuable, something that has been associated with Britain for generations – they have come to symbolise the simple principle of individual freedom.

On July 18th 2003 Prime Minster Tony Blair, the man who in 1999 promised an anti-hunting Bill, justified the war with Iraq to the American Congress, with the words: 'We are fighting for the inalienatable right of human kind .... To be free. Free to be you, so long as being you does not impair the freedom of others. That's what we're fighting for and that's a battle worth fighting for.' It is a great pity that he does not say the same thing – and mean it – when in Britain.

So what should country people and country sportsman do as their freedoms are attacked by unscrupulous politicians – is civil disobedience one answer? Perhaps they should look to the writings of an American – Mark Twain – in his *A Yankee at the Court of King Arthur*; and make up their own minds.

'You see my kind of loyalty was loyalty to one's country, not to its institutions or its office-holders. The country is the real thing, the substantial thing, the eternal thing; it is the thing to watch over and care for, and be loyal to; institutions are extraneous, they are its mere clothing, and clothing can wear out, become ragged, cease to be comfortable, cease to protect the body from winter, disease and death.'

Sadly, the institution of Parliament has become worn out and country people are not receiving the protection and freedom under the law they deserve.

OPPOSITE: *A painting by John King.*

PAGES 186–7: *'The Spectators' by Rodger McPhail*

R. McPhail

As the bees gather their honey from the broad stretches of heather, so those who go out into the open air gather up vigour of frame and infinite nervepower which is more valuable than muscular strength.

Nothing but sport can supply it, and thus the country has a value over and above its utilitarian produce. A moor – a vast stretch of heather – may graze a few sheep: the money they represent is but little. But the grouse give an increase of strength, a renewal of nerve-focus, to those who pursue them over the mountain side, not to be estimated in pounds, shillings and pence. A little trout stream, if it were farmed on the most utilitarian principle, could only send a small tribute of fish towards feeding a town. But the same river may lead many and many a sportsman out into the meadows, insensibly absorbing the influence of the air and sunlight, the wood and hills, to his own profit individually and to the benefit of all with whom he associates.

A fox is useless in itself, but each fox fairly hunted is worth a thousand pounds: a thousand pounds worth of health, courage, manliness, and good-fellowship are purchased by a successful run. These are things of the very highest value, not only to the individual but to the country.

Richard Jefferies: *A Defence of Sport* 1883

*

As an interglacial man, I feel no embarrassment, except for one thing – that we ended the hunting way. It had shaped us, given us – anatomically and socially – the way we are. But we killed off our fellow species in the natural world. The death of the hunter and the hunted must be the sin that interglacial man committed in the memories of his inheritors. How do you live when the tundra returns but not the reindeer, the aurochs, the extinct mammoth?

Robert Ardrey: *The Hunting Hypothesis* 1976

*

Western Europe was once governed by a country mind – even city states had a keen country awareness – but the country mind has greatly diminished. People who know nothing about the land suddenly sit and rule like gods over the fate of other men.

Sir Laurens van der Post: *A Walk with a White Bushman* 1986

ACKNOWLEDGEMENTS

Without the help of many people, it would have been impossible to write both the original edition and the second edition of this book. First and most importantly to my parents for looking after my dog, often at short notice, to allow me to travel to various parts of the country. Also to my brother, John, on the farm, for accepting my unreliability with his usual good nature. My sister, Rachael, sister-in-law, Ellen and Margaret too, for checking the manuscript and to Teresa Brown for typing the final draft.

Many real experts in their field helped with information or reading the text – Dr David Hill, Dr Peter Hudson, Dr Dick Potts, Dr Stephen Tapper, Dr Derek Mills, Mr Richard Prior and Mr John Hotchkis.

Many organisations also gave me valuable help: The Nature Conservancy Council, the Royal Society for the Protection of Birds, the Wildfowl Trust, the Royal Society for Nature Conservation, the British Field Sports Society, the British Association for Shooting and Conservation, the Red Deer Commission, the British Deer Society, the Atlantic Salmon Trust, the Scottish Office, the Department of the Environment, the Game Conservancy, the Anglian Water Authority, the Forestry Commission, the Atlantic Salmon Conservation Trust, the Anglers' Co-operative Association, the National Federation of Anglers, the National Anglers' Council, the Swan Rescue Service (Europe), the Otter Trust, and the Farming and Wildlife Advisory Group.

The following individuals (sadly many now deceased since the first edition was published) – members of organisations, landowners and others gave much assistance during the preparation of the original book; Eric Carter, Juliet Keith, Arthur Illes, Brian Sewell, Sir Stephen Hastings, Capt Ronnie Wallace, Lt Col Tony Murray-Smith, 'Pop' Payne, Joan Wood, Jim Hunter, Noel Cunningham-Reid, Tom Reeve, Sir Mark Prescott, Fenton Kirwan, Stewart Willcock, John Wilson, Christopher Pope, John Hotchkis, Hugh Oliver-Bellasis, Christopher Passmore, John Grant, Tilly and Alan Smith, Norman McCulloch, Michael Chinery, the Earl Peel, Richard Porter, Gordon Beningfield, Will Garfit, Bruce Brown, Len Baker, Malcolm Lyell, Peter Weston, Peter Tombleson, Philip Wayre, Dr Gareth Thomas, Jeremy Sorensen, Cliff Carson, Don Revett, Richard Van Oss, John Anderton, John Swift, Dr Peter Mayhew, Dr Colin Shedden, Tony Laws, Michael Street, Hugo Wood-Homer, John Pugsley, John Hitchings, Nick Fox, Lord Home of the Hirsel, Rear Admiral John Mackenzie and the Hon James Stuart.

I am particularly grateful to the late Sir Laurens van der Post and 'BB' (the late Denys Watkins-Pitchford) for permission to quote from their work and to 'BB' for allowing the reproduction of two of his pictures. Artists Rodger McPhail, Will Garfit, Gordon Beningfield and John King were kind enough to permit some of their work to be used. Will Garfit specially drew the excellent chapter title pages and jacket, and John King embellished the Fox chapter and others. The *Daily Telegraph* also kindly allowed me to incorporate two of my articles, on hare and salmon, into the text.

Finally I would like to thank Reggie Lofthouse, the Convener for the Standing Conference on Countryside Sports, and the late Lord Porchester, the Chairman, for their help and encouragement. For any person or organisation that I have omitted, I do apologise – it signifies a lapse of memory, not a lack of appreciation.

Illustration Credits

Heather Angel: pp 37, 71, 145, 146, 168
Arthur Ackermann & Son Ltd.: pp 26, 28, 101, 148, 171
'BB' – D.J. Watkins-Pitchford: pp 62, 175
Gordon Beningfield: p 18
Thomas Bewick: pp 24, 104, 128, 131, 150
The British Association for Shooting and Conservation: pp 164 (bottom left & right)
The British Deer Society: pp 78, 124, 125, 159
The British Field Sports Society: p 31
Neil Cook: pp 61, 82, 121, 122, 127
Frank Lane Picture Agency: pp 17, 23, 33, 56, 136
The Game Conservancy Trust: pp 58, 77, 84, 96, 102 (both), 118 (both)
William Garfit: pp 11, 25, 49, 67, 83, 99, 115, 129, 132, 151, 158, 161, 177
John King: pp 34, 39, 40, 43, 48, 51, 55, 72, 185
Rodger McPhail: pp 75, 103, 119, 155, 176, 186/7
Rodger McPhail commissioned by Holland & Holland for the engraving of The Set of Four British Field Sports Guns: p 3
Robin Page: pp 2, 13, 14, 29, 41, 44, 45, 53, 69, 87, 90, 106, 107 (both), 138, 141, 142, 164 (top)
The Parker Gallery: pp 64, 79, 114, 133, 137(both), 162, 172
Christopher Passmore: pp 111, 112
Dr. Franklyn Perring: p 21
Phillipa Scoones: p 95
Fiona Silver: pp 70, 98 (both), 134